Transition Pathways towards Sustainability in Agriculture: Case Studies from Europe

Transition Pathways towards Sustainability in Agriculture: Case Studies from Europe

Lee-Ann Sutherland. *James Hutton Institute, Aberdeen*

Ika Darnhofer. *University of Natural Resources and Life Sciences, Vienna*

Geoff A. Wilson. *Plymouth University, Plymouth*

Lukas Zagata. *Czech University of Life Sciences, Prague*

www.cabi.org

CABI is a trading name of CAB International

CABI
Nosworthy Way
Wallingford
Oxfordshire OX10 8DE
UK

Tel: +44 (0)1491 832111
Fax: +44 (0)1491 833508
E-mail: info@cabi.org
Website: www.cabi.org

CABI
38 Chauncy Street
Suite 1002
Boston, MA 02111
USA

T: +1 800 552 3083 (toll free)
E-mail: cabi-nao@cabi.org

A catalogue record for this book is available from the British Library, London, UK.

Library of Congress Cataloging-in-Publication Data

Sutherland, Lee-Ann.
Transition pathways towards sustainability in agriculture : case studies from Europe / Lee-Ann Sutherland, James Hutton Institute, Aberdeen, Ika Darnhofer, University of Natural Resources and Life Sciences, Vienna, Geoff A. Wilson, Plymouth University, Plymouth, Lukas Zagata, Czech University of Life Sciences, Prague.
pages cm
Includes bibliographical references and index.
ISBN 978-1-78064-219-2 (hbk : alk. paper) 1. Sustainable agriculture--Europe--Case studies. 2. Agricultural systems--Europe--Case studies. I. Title.

S452.S88 2014
338.1094--dc23

2014033143

ISBN-13: 978 1 78064 219 2

Commissioning editor: Claire Parfitt
Production editor: Claire Sissen

Printed and bound in the UK from copy supplied by the authors by CPI Group (UK) Ltd, Croydon, CR0 4YY.

Contents

Figures and Tables

Figures

Tables

Contributors

Dr Ika Darnhofer. Department of Economic and Social Sciences, University of Natural Resources and Life Sciences (BOKU), Vienna, Feistmantelstr. 4, 1180 Vienna, Austria.

Dr Catherine Darrot. AGROCAMPUS OUEST. Laboratoire de Développement Rural, UMR ESO 65 rue de Saint Brieuc. CS 84215, 35042 Rennes, France.

Ing Marion Diaz. AGROCAMPUS OUEST. Laboratoire de Développement Rural, UMR ESO 65 rue de Saint Brieuc. CS 84215, 35042 Rennes, France.

Dr Mariana Draganova. Institute for the Study of Societies and Knowledge, 13-A Moskovska St., 1000-Sofia, Bulgaria.

Dr Sharon Flanigan. Social, Economic and Geographic Sciences Group, James Hutton Institute, Craigiebuckler, Aberdeen, AB15 8QH, United Kingdom.

Dr Carla Gonzalez. Institute for Mediterranean Agrarian and Environmental Sciences (ICCAM), Universidade de Évora, Pólo da Mitra, Edifício Principal, Gabinete 202, Apartado 94, 7002-554 Évora, Portugal.

Dr Helena Guimarães. Institute for Mediterranean Agrarian and Environmental Sciences (ICCAM), University of Évora, Polo da Mitra, 7000 Évora, Portugal.

Dr Pavlos Karanikolas. Department of Agricultural Economics and Rural Development, Agricultural University of Athens, Iera Odos 75, 11855 Athens, Greece.

Dr Michal Lošťák. Faculty of Economics and Management, Czech University of Life Sciences Prague, Kamycka 129, 165 21 Prague, Czech Republic.

Dr Annie McKee. Social, Economic and Geographic Sciences Group, James Hutton Institute, Craigiebuckler, Aberdeen, AB15 8QH, United Kingdom.

Prof Plamen Mishev. Department of Natural Resources Economics, University of National and World Economy (UNWE), Student town 'Hristo Botev', Sofia 1700, Bulgaria.

Dr Mariya Peneva. Department of Natural Resources Economics, University of National and World Economy (UNWE), Student town 'Hristo Botev', Sofia 1700, Bulgaria.

Ms Sarah Peter. Institute for Rural Development Research (IfLS), Kurfürstenstr. 49, 60486, Frankfurt am Main, Germany.

Dr Teresa Pinto-Correia. Institute for Mediterranean Agrarian and Environmental Sciences (ICCAM), University of Évora, Polo da Mitra, 7000 Évora, Portugal.

Ms Simone Schiller. Institute for Rural Development Research (IfLS), Kurfürstenstr. 49, 60486, Frankfurt am Main, Germany.

Prof Bill Slee. Social, Economic and Geographical Sciences Group, James Hutton Institute, Craigiebuckler, Aberdeen, AB15 8QH, United Kingdom.

Dr Lee-Ann Sutherland. Social, Economic and Geographic Sciences Group, James Hutton Institute, Craigiebuckler, Aberdeen, AB15 8QH, United Kingdom.

Ms Emi Tsakalou. Department of Agricultural Economics and Rural Development, Agricultural University of Athens, Iera Odos 75, 11855 Athens, Greece.

Dr George Vlahos. Department of Agricultural Economics and Rural Development, Agricultural University of Athens, Iera Odos 75, 11855 Athens, Greece.

Prof Geoff Wilson. School of Geography, Earth and Environmental Sciences, University of Plymouth, Drake Circus, Plymouth, PL4 8AA, United Kingdom.

Dr Lukas Zagata. Faculty of Economics and Management, Czech University of Life Sciences, Kamycka 129, 165 21 Prague, Czech Republic.

Glossary

AAFNs	Alternative agri-food networks
ADPM	Association for Mértola Heritage Protection
AGRO2	Standards for the Inspection of Integrated Crop Management Systems
AGROCERT	Agricultural Products Certification and Supervision Organization
ANKA	Anka Development Agency of Karditsa S.A.
BSPB	Bulgarian Society for the Protection of Birds
CAP	Common Agricultural Policy
CEDAPA	Centre d'Etude pour un Développement Agricole Plus Autonome
CHP	Combined heat and power
CMO	Common Market Organisation
CO_2	Carbon dioxide
CPA	Local Professional Committee of Agriculture
CRIE Montado	Centro Regional de Inovação do Montado
CZK	Czech Republic Koruna
EAFRD	European Agricultural Fund for Rural Development
EC	European Commission
EEC	European Economic Community
EEG	Erneuerbare-Energien-Gesetz
EU	European Union
EU-27	The 27 Member States of the European Union
FarmPath	Farming Transitions: Pathways towards regional sustainability of agriculture in Europe (project)
FiTs	Feed-in Tariffs
FRCIVAM	Regional Federation of the Initiative Centres to Valorize Farming and Rural Areas
GATT	General Agreement on Tariffs and Trade
GDP	Gross Domestic Product
GMOs	Genetically Modified Organisms
HNV	High nature value
HNVF	High nature value farming
IAD	Institutional analysis and development
ICM	Integrated crop management
IT	Information technology
LAGs	Local Action Groups
LEADER	Liaison Entre Actions de Développement de l'Économie Rurale

LFA	Less Favoured Area
MLP	Multi-level perspective
MULTAGRI	The Multifunctionality of Agriculture and Rural Areas (project)
NGO	Non-governmental organization
NSPG	National Stakeholder Partnership Groups
NUTS	Nomenclature of Territorial Units for Statistics
OECD	The Organisation for Economic Co-operation and Development
PDO	Protected Destination of Origin
PGI	Protected Geographical Indication
PROVERE	Program for the Economic Value of Endogenous Resources
RDP	Rural Development Programme
RES	Renewable energy source
RNP	Regional Nature Park
ROCs	Renewable Obligation Certificates
ROAAS	Regional Office of the Agriculture Advisory Service
SA	Sustainability assessment
SFP	Single Farm Payment
SOLINSA	Support of Learning and Innovation Networks for Sustainable Agriculture
SRDP	Scottish Rural Development Programme
SWM	Stadtwerke München GmbH
TOPMARD	Towards a Policy Model of Multifunctional Agriculture and Rural Development Research (project)
UAA	Utilised Agricultural Area
UK	United Kingdom
UNESCO	The United Nations Organization for Education, Science and Culture
USA	United States of America
WTO	World Trade Organization
YF	Young Farmers

Acknowledgements

The chapters in this book are based on findings from the FarmPath research project (Farming Transitions: Pathways towards regional sustainability of agriculture in Europe) (Grant agreement no 265 394), undertaken from 2011 to 2014 (www.farmpath.eu). The project was initiated in response to a call from the European Commission's 7th Framework Programme, which sought projects under the topic 'Assessment of transition pathways to sustainable agriculture and social and technological innovation needs'. Approximately 72% of the funding for the research was received through the European Commission, with the remainder provided by the participating institutions:

- James Hutton Institute, Aberdeen, United Kingdom
- University of Natural Resources and Life Sciences, Vienna, Austria
- Agricultural University of Athens, Greece
- University of National and World Economy, Sofia, Bulgaria
- Institute for Rural Development Research, Frankfurt am Main, Germany
- Czech University of Life Sciences, Prague, Czech Republic
- University of Évora, Portugal
- Institute for Life, Food and Horticultural Sciences and Landscaping, Rennes, France
- Plymouth University, United Kingdom

In total, we estimate that over 1,000 people from across Europe were involved in the project, as members of National Stakeholder Partnership Groups, interviewees, focus group and workshop participants, researchers, and academic advisors. We wish to recognize our international advisory group: Dr Boelie Elzen (University of Twente), Prof Otto Schmid (FiBL), Prof Hilkka Vihinen (MTT), Ingrid Petterson (CEJA), and Maciej Krzysztofowicz (European Commission) for their feedback and advice on project activities and outputs. We are particularly grateful to Dr Claire Kelly and Dr Nichola Harmer (Plymouth University), for their editorial assistance: revising the English on several chapters, preparing figures, tables and the glossary, cross-checking references, and compiling the chapters into a consistently formatted whole. We are also grateful to Tim Absalom (Plymouth University) for producing the case study location maps.

This book is dedicated to our families,

who enable us to do the work we love.

Chapter 1

Introduction

L-A. Sutherland[1], G.A. Wilson[2], L. Zagata[3]

[1]*James Hutton Institute, Aberdeen (lee-ann.sutherland@hutton.ac.uk);* [2]*Plymouth University;* [3]*Czech University of Life Sciences, Prague*

What this book is about

Over the past decade the transition towards sustainable agriculture has been a central theme in the work of governments, NGOs and research institutions. Multiple publications, including the European Commission White Paper on Adapting to Climate Change (2009) and various academic publications (e.g. Wilson, 2007; Brouwer and van der Heide, 2009), identify the importance of increasing sustainability of agriculture in order to meet future challenges. However, despite the adoption of the notion of sustainable development of agriculture as a basic EC policy principle (see Council Regulation (EC) No1257/1999), it is becoming clear that changes are needed to ensure that agriculture in the EU can meet the increasing range of public goods and functions desired by citizens. For example:

- Natural resource use (land, water) is increasingly contested by non-agricultural uses (e.g. fibre and energy) (EEA, 2006a, 2006b, 2006c). Farming itself is highly dependent on non-renewable energy resources (Heinberg and Bomford, 2009).
- Decreasing support for farm production and encouragement of commercial competitiveness coincide with increased national and EU public regulatory intervention in farming (Robinson, 2008).
- Production and marketing chains are becoming more concentrated, both upstream (input provision) and downstream (e.g. output distribution, high level of concentration in the food retail market) (Campbell, 2005; Potter and Tilzey, 2005, 2007). At the same time, there is a growing range of 'alternative' products, production and retail options, challenging these globalizing tendencies (van der Ploeg, 2007).
- Consumers and citizens are demanding higher standards of food safety and quality control. Concerns exist surrounding the nutritional value of highly processed food and the social and ecological impacts of purchasing decisions (e.g. Fair Trade, food miles) (Lang and Heasman, 2004).
- A sharp rise in food prices from 2007 to 2008 is challenging two decades of 'non-productivist compromise' in agricultural policy in Europe, with food security re-entering the political agenda (Wilson, 2007; Marsden, 2013).

- Of particular concern has been the potential impact of climate change, especially with regard to water availability (mainly in southern Europe, but also in northern and Eastern Europe) and shifting agricultural zones northwards, leading to increased environmental uncertainty for rural stakeholders (Cline, 2007; Mestre-Sanchis and Feijoo-Bello, 2009) and potential impacts on production (Wilson, 2007).

In the European context, issues linked to rural development (e.g. depopulation and ageing in rural areas, preservation of cultural landscapes, demand for recreation spaces close to urban areas) also play an important role in the search for a more sustainable agri-food system (Robinson, 2008). There is also concern about the lack of young people in agriculture. Some European countries suffer from a shortage of young people taking up farming as a profession, caused by inadequate generational turnover and leading to ageing populations and greying communities (Eurostat, 2011).

These challenges demonstrate both the need and opportunity for transition in European agriculture. In this book we assess transition processes: fundamental changes that incorporate processes of societal, ecological, economic, cultural, technological and institutional co-evolution (Loorbach and Frantzeskaki, 2009). Transitions involve several sectors or sub-sectors as well as a range of societal actors at multiple scales. Through the interdependency and co-evolution of these, society or an important societal subsystem, fundamentally changes. A transition is thus qualitatively different from an incremental change that is limited in scope (e.g. does not affect a whole sector of the economy), in time (is only a fad and does not stabilize) or in space (only takes place in some regions). For a transition to occur, different developments and actors at regional and national levels have to come together, engendering a development pathway based on new practices, technologies, knowledge, institutions, social organization, guiding principles and values.

The aim of this book is to improve our understanding of transition processes in European agriculture. We focus on 'emerging transitions' – how new organizational forms and technologies change, and are changed by, mainstream actors and practices in the agricultural sector. This is achieved through an integration of recent academic theory on transition and change in agricultural systems and assessing its utility for empirical research. The multi-level perspective (MLP) on system transition (Geels *et al.*, 2004; Geels and Schot, 2010) is applied to clusters of case studies, which focus on different types and aspects of transition processes within agriculture. Our purpose in studying transition in agriculture is not only to understand how change has occurred, but to assess how intervention can successfully be made to facilitate sustainability transitions. Most of the cases studied in this book have had active policy intervention, although some demonstrate that transition processes can occur endogenously without such assistance. In this chapter we delve further into the definitions of sustainability and transition, key challenges to achieving sustainability transitions, and conclude with brief insights into the content of subsequent chapters.

Sustainability in agriculture

Sustainability challenges in European agriculture: recession, outmigration, global competition

The definition of sustainable agriculture – and indeed whether this is a meaningful goal – is highly contested in the literature on agriculture and rural development (Robinson, 2008). However, it is widely agreed that increasing the sustainability of agricultural systems is a necessary and important objective (Pretty, 2002). Increasing the sustainability of agriculture and agricultural systems is a long standing goal of European agricultural policy (Marsden, 2003; Wilson, 2007; Robinson, 2008). The sustainability of European agriculture can be broadly conceptualized along three main processes: economic, environmental and social. These three processes provide an important conceptual basis for the framing of the arguments developed in this book.

At a time of global recession, issues surrounding the *economic sustainability* of European agriculture have yet again come to the fore. The post-2007 recession has exacerbated economic problems faced by almost all European countries, in particular by amplifying the widening income gap between rural and urban areas, by further marginalizing non-globalized farming areas in uplands or 'remote' parts of Europe, and by reducing opportunities for on-farm multifunctional activities such as farm tourism or specialist food sales (Wilson, 2010). With its rapid onset and resulting impact on the availability of funds provided by banks and lending institutions, the economic recession can be conceptualized as a sudden shock, in that it has catapulted many agricultural systems towards lower sustainability in a relatively short time span. However, the most important aspects of the recession have been the ripple effects caused by sudden changes in the economic climate. In the context of our European case studies, these have included, in particular, reduced demand for agricultural products and services (i.e. the sale of high-end organic and/or locality-specific food products), reduced borrowing opportunities for upgrading or purchasing farm equipment and buildings and, most importantly, reduced available funds for young farmers wishing to take up farming (e.g. mortgages, finance for land purchase). The impact of the recession on tourism in European rural areas has been particularly pronounced, although early evidence suggests a geographically varied picture. While some regions have reported falling tourist numbers and reduced income for rural stakeholders, other areas appear to have benefitted from the fact that tighter household budgets have meant fewer trips abroad, with a concurrent increase in domestic tourism. In the United Kingdom (UK), areas such as the south-west or Scotland, for example, appear to have weathered the recession relatively well with regard to tourism figures. Overall, the recession has increased uncertainty surrounding the day-to-day planning of farm activities, has reduced incentives for multifunctional endeavours that tend to raise sustainability of farming systems, and has severely affected rural communities that were already in the process of 'losing touch' with their agricultural stakeholders. While many of these processes will not be evident for several years, our case studies will show that many of the warning signs associated with declining socio-economic capital are already becoming visible.

The post-2007 global recession has also exacerbated trends towards rural outmigration, land abandonment (especially in southern Europe, but also in remote upland areas), decreasing family farm succession, and a relative loss of the former hegemonic position of agriculture in European societies. Land abandonment, in particular, has led to substantial environmental challenges where insufficient labour remains to maintain fragile and work-intensive structures such as dry stone terraces, complex irrigation systems or farm woodlands. The relative economic marginalization of agriculture in Europe (accounting for only 3 to 6% of GDP on average, excluding food processing) has been paralleled by an increasing urban bias, with a tendency for funds, planning effort and stakeholder support to be focused on improving livelihoods for urban rather than rural populations (Woods, 2005).

In parallel, increased competition related to globalization processes, the gradual embeddedness of even the remotest European agricultural areas into global markets (including new Eastern European EU member states) and the associated loss of localized and often sustainable food production systems, have further affected the economic viability of many rural areas. While pro-globalization commentators have suggested that increasing globalization can lead to major economic advantages for rural areas (e.g. improved access to funds and knowledge), anti-globalization proponents have highlighted how global embeddedness often weakens local economic capital by crowding out small locally-based producers, by creating vertical economic ties that weaken horizontal (within and between communities) embeddedness, and by devaluing local production and quality (Gray, 2002; Stiglitz, 2002; Woods, 2005). Transitions engendered by globalization often lead to an increasing divide between wealthy and poor stakeholders in village communities, whereby the 'winners' of globalization (i.e. those who are better vertically networked) are often able to accumulate wealth that rarely trickles down to those who are less well networked (Woods, 2005; Wilson, 2012). Globalization also leads to substantial changes in economic activities in rural areas, most frequently away from low-input agricultural production towards profit-driven services. This, in turn, often leads to further outmigration or to a situation where potential farm successors must seek off-farm employment in the local area.

Globalization has also often been reported to change the socio-psychological setup of rural areas. Indeed, formerly close-knit communities focused almost entirely on agricultural production (that often required strong networks of trust and assistance) have now increasingly become 'hybrid' communities with multiple economic pathways, of which agriculture is only one of many. The outcome has often been a feeling of disassociation between residents and communities, loss of community cohesion, and ultimately, loss of community resilience (Wilson, 2010, 2012). Globalization – and associated policies linked to the World Trade Organization – has increasingly allowed global competition to affect what were traditionally local agricultural markets. Thus, today's rural areas in Europe not only have to face strong internal competition but are also increasingly exposed to global market forces, reduction or abolition of previously protected agricultural markets, and lower prices for agricultural goods which can be produced more cheaply in export-oriented non-EU countries. As Potter and Tilzey (2005) argued, this has in many ways highlighted the weaknesses of the 'European model' of agriculture based on highly subsidized, but internationally uncompetitive, family farms. With the planned demise of direct agri-environmental payments to European farmers, and the increasing pressure exerted by global agricultural exporters such as the USA on European agricultural markets, for many European farming regions it is only a question of time as to when farming will become unprofitable – leading to a downward spiral of further farm abandonment and outmigration. The case studies in this book have been selected specifically to include more, and less, globalized areas of Europe and a key question will be to what extent the global connectedness of these areas has affected sustainability transitions on the ground.

The acceleration of outmigration from rural areas, linked to globalization, addresses issues of *social sustainability*. In both the agricultural literature and the general discourse on sustainable development, the social pillar of sustainability is conceptualized flexibly. Boström argues that the meaning of the term 'social sustainability' is not clear and that "there exists uncertainty about how it relates to both the other dimensions and wider policy issues" (Boström, 2012:7). Authors reviewing research literature and strategic policy documents show a myriad of themes covered by the social sustainability concept, spanning equity, poverty and social inclusion, health and education, population dynamics and governance (Murphy, 2012). A similar range of topics can be seen in discussions focused specifically on sustainability of agriculture. European visions in this area emphasize a balance between economic, environmental and social dimensions, and demonstrate the scale of the issue of social sustainability within agriculture, such as production of high quality food, adequate income for farms, quality life in rural areas, or balanced territorial development (European Commission, 2012). Above all, social sustainability of agriculture can be monitored across different spatial dimensions (regional or national) and at various temporal scales (long- or short-term).

Outmigration has been one of the key factors reducing the sustainability of rural systems ever since mass migration to rapidly growing urban centres began in the UK in the late 18th century. However, evidence throughout Europe shows that rural outmigration has accelerated substantially and that rural migration patterns are temporally and spatially complex. In many areas, especially in the peri-urban fringe or near key transport nodes, populations have substantially increased due to processes of counter-urbanization and have also begun to affect rural areas in post-communist Eastern Europe (especially in the past 10 years). In countries such as the UK where counter-urbanization became dominant from the 1970s onwards, house price inflation in rural areas has begun to prompt movement back from the countryside towards urban areas (Woods, 2005). However, although helping to maintain populations and, to some extent, local services, counter-urbanized communities have contributed further towards transitions *away* from agriculture, as traditional agricultural stakeholder groups have often been replaced by middle class urban migrants with little or no connection to the land.

In many European rural areas, outmigration is nonetheless still a real threat, severely affecting the sustainability and viability of rural communities (Wilson, 2010). Again, processes are complex and often region-specific, but overall there has been a tendency for young people to leave their villages, leading to 'greying' communities marred by loss of services, lack of labour to tend agricultural land, and leading to substantial problems of recruiting farm successors. As highlighted above, the loss of people to help with family farming businesses often leads to loss of natural capital, especially in labour-intensive agricultural systems that rely on the maintenance of landesque capital (irrigation channels, terraces, etc.). Indeed, land abandonment has become an unfortunate feature of European rural communities, with associated transition of former agricultural landscapes into forest, shrubland or, most worryingly, desertified areas in which soil productivity is rapidly lost (Imeson, 2012).

Our conceptual foundation of the social sustainability of agriculture emphasizes the role and importance of young people with regard to their potential to contribute to transition processes towards a more sustainable model of agriculture in Europe. The main purpose of this concept is to directly explore the problem of ageing farming populations, considered from a policy perspective as a key threat to the future sustainability of agriculture in Europe (DGIP, 2012). A brief review of the statistics suggests that there is a 'young farmer problem' in Europe: almost one third of farms have a land-holder who is above 65 years of age (Eurostat, 2011), while young farmers (< 35 years) can be found only on 7.5% of farms in Europe. Yet a closer look at the 'young farmer problem' opens additional questions that challenge some of its core arguments. First, there are considerable differences between countries, and age structures often correspond with farm-size structures. Second, there is insufficient empirical evidence to explain at what level the shortage of young people in agriculture becomes a problem, and what the negative consequences of outmigration may be. Regarding social sustainability, the most worrying situation is in countries that have a shortage of young farmers and, at the same time, have a relatively high proportion of older farmers (Zagata and Lošťák, 2014). Yet negative consequences are not related to the societal aspects of agriculture alone, but interact with others, especially economic and environmental processes. The available evidence shows that enterprises with young farmers tend to generate higher income and are more productive (Eurostat, 2011). Some of the case studies presented in this book have therefore looked for evidence of why different farming models appeal to young people and new entrants to farming, and what obstacles they face in taking up farming as a career and/or lifestyle choice.

The *environmental sustainability* of European agricultural areas has faced similar challenges. The post-war era has been characterized by a transition from productivism to non-productivism, with increasing emphasis on environmental conservation of the countryside particularly evident from the mid-1980s onwards (Wilson, 2001, 2007; Marsden, 2013). While the era of productivism was seen generally as a time of maximization of agricultural productivity and income, with little regard for the environment, the non-productivist era has been characterized by an emphasis on multifunctional use of the countryside which includes both agricultural production and environmental conservation (Wilson, 2007; Robinson, 2008). However, as various commentators have emphasized (e.g. Marsden *et al.*, 1993; Wilson, 2001), this transition has been spatially and temporally uneven, with some lowland and fertile European agricultural areas becoming ever more productivist (so-called 'super-productivism'), while upland and marginal agricultural areas have increasingly taken non-productivist pathways characterized by low-intensity and (usually) environmentally-friendly farming (Wilson, 2010). It is the latter areas that have particularly benefited from agri-environmental subsidies, although there continues to be considerable debate as to whether agri-environmental payments have led to a genuine improvement in the environment or just acted as 'green subsidies' to keep economically marginal farm households 'on the land' (Potter and Tilzey, 2005; Dibden and Cocklin, 2009).

Most worryingly from an environmental sustainability perspective have been developments over the last 10 to 15 years which suggest that the transition towards a more environmentally sustainable countryside has been only partial at best. Whilst the 1990s could be regarded as the 'greenest' decade in European agriculture with regard to policy support (e.g. up to 20% of EU farmland under set-aside and >25% of agricultural land under agri-environmental payments, Buller *et al.*, 2000), the 2010s saw an overall reduction in support for low intensity agriculture despite much policy rhetoric related to the CAP 2nd

pillar of support (Robinson, 2008; Brouwer and van der Heide, 2009). The reasons for this are manifold. First, increasing global demands due to the rapidly changing diets of almost half of the world's population (especially India, China and other transition economies) have led to intensive dairy and meat production which leaves little room for agricultural extensification. Indeed, this is one of the key drivers for the development of environmentally harmful super-productivist agricultural pathways geared almost entirely towards global agri-commodity chains (Robinson, 2008). This change has been exacerbated by the use of agricultural land for production of biofuels, which has taken more agricultural land away from production, thereby increasing pressure on the remaining land for food production. Second, as Potter and Tilzey (2005) have amply demonstrated, increased pressure from the World Trade Organization and increasing pressure to open European agricultural markets to global trade (especially from the USA) has put additional pressure on rural areas in Europe, with the fragile 'family farm' model becoming an anachronism in an increasingly globalized and neo-liberal world (Wilson, 2007). Yet it is the low-intensity family farm that has often been praised as the model for an environmentally sustainable and locally well embedded farm. Third, the global population is increasing by c.75 million people per year, meaning that an additional population of equivalent size to that of Germany needs to be fed every year. European agriculture must therefore play its part in increasing production to satisfy rapidly changing food needs. Fourth, these processes are exacerbated by the threat of climate change which not only limits the capacity of some European agricultural areas to increase production (through limited water availability, for example) but may also change what can be grown in certain areas (Cline, 2007; Mestre-Sanchis and Feijoo-Bello, 2009). In combination, these processes may suggest that super-productivist (or indeed neo-productivist) pathways of agricultural change may begin to dominate, leaving less and less room for more environmentally sustainable non-productivist areas in many European regions – an issue that lies at the heart of the European case studies presented in this book.

Sustainability as a regionally defined concept

In light of the number of potential functions and challenges facing agriculture in Europe, we propose that increasing sustainability is best addressed by enabling flexible combinations of farming models, which vary to reflect the specific opportunity sets embedded in regional culture, agricultural capability, diversification potential, ecology, historic ownership, and governance structures. The definition of sustainability at regional level reflects a shift away from the notion that individual farms or farming systems can or should be expected to meet the full range of public and industry demands on agriculture. Instead, we propose that these demands should be met at regional[1] level, through flexible combinations of approaches to farming. The notion that sustainability is achieved at regional level not only takes into account that there will be regional differences in the forms

[1] We acknowledge that the definition of 'region' is highly contested: the NUTS (Nomenclature of Territorial Units for Statistics) subdivisions, while an important distinction for the delivery of EU structural funds, do not necessarily adhere to geographic or administrative boundaries. In terms of scale our definition of 'region' is at NUTS 3 or 4 classification levels (i.e. sub-national divisions). In practice, transition towards regionally sustainable agriculture could be expected to involve increasing the authority of regional governments over agricultural funding and intervention.

and capabilities of agriculture, but also includes the tenet that interactions between individual farm models and farming systems at regional level are a key aspect of sustainability. Different approaches to farming can be expected to provide different public goods, functions and ecosystem services (e.g. food security, employment, public access, energy crops), and they are expected to interact in complementary as well as conflicting ways (e.g. local food competing with organic food on supermarket shelves). However, what is crucial is whether farming models present within a region interact in such a way as to meet the changing needs and demands of consumers, members of the production system, and citizens (both within the region and more broadly across Europe), while also providing socially and economically viable livelihoods for farm households. Sustainability of agricultural systems is, thus, an ongoing, adaptive process of enabling farming households and members of agricultural production and consumption chains to respond to the changing needs and preferences of consumers and citizens, through flexible combinations of farming models, and provision of a suite of public goods and agricultural functions at regional level. It is a process, rather than an end goal.

The notion that multiple functions and public goods could and should be provided through agricultural systems is a foundational tenet of EU agricultural policy (van Huylenbroek and Durand, 2003). A considerable volume of academic literature over the past decade has identified the multiple functions provided by agriculture as key to sustainability in agricultural systems (see in particular van Huylenbroek and Durand, 2003; Brouwer and van der Heide, 2009 for reviews of the multifunctional agriculture literature). Indeed, this was the initial premise of both the TOPMARD and MULTAGRI EU Framework 6 projects, as well as conceptual work by Wilson (2007, 2008). However, to date analyses of multifunctionality in agriculture have focused primarily at the macro level, on policies developed to promote or support multiple functions in agriculture (Potter and Burney, 2002; Brouwer, 2004; Potter and Tilzey, 2005), with some work at farm level, to evaluate the degree to which multifunctional transitions are occurring on the ground (Wilson, 2007, 2008).

In line with Clark (2005, 2006) we suggest that the multiple functions and public goods provided through agricultural systems should be assessed and enabled at regional level. In light of the increasing demands on agriculture, we argue that it is not reasonable to expect each individual farm or farming system to attempt to meet all of these demands. Neither is it acceptable to have entire regions where some of these demands – such as public access, local food production or maintenance of soil and water quality – are not met, as might be the case if public goods and multiple functions were pursued primarily at national or European levels. Instead, we propose that sustainability can best be addressed by enabling regions to optimize the specific opportunity sets embedded in those regions (e.g. natural resources, farm structures and farming systems, landscape and ecological conditions, infrastructure and economic development, social capital, institutional arrangements, governance structures). Linked to this, the particular issues of sustainability to be addressed will vary between regions and through time within the same region. For example, in some regions productivist agriculture is still the dominant activity, where issues of sustainability relate to increasing farm size and ecological degradation resulting from intensive production. In other regions, more commonly remote or peripheral regions, other activities such as hunting or nature related recreation/tourism are already taking over as management goals for some farm units. In yet other regions, subsistence farming, as well as abandonment of farming and of the land area in general, are the dominant processes (Wilson, 2001). Even within a region, challenges to economic, social and environmental sustainability are bound to vary through time, for example, as new scientific evidence becomes available, as markets shift or as societal awareness and expectations evolve. As Robinson (2008) has highlighted, definitions of sustainability of agriculture are, therefore, inevitably region specific as are the pathways to achieve economic, ecological and social sustainability. The research in this book will provide an evidence-base to support this definition of sustainability and identify mechanisms to enable transitions towards achieving it.

Conceptualizing emerging transition processes

Over the last two decades, a number of theoretical and methodological perspectives associated with attempts to enable transition to more sustainable agriculture are discernible. The more distant roots lie in mono-disciplinary studies which often highlighted problems of water quality and quantity, declining biodiversity, landscape change and land fragmentation, but which have rarely effected transformative change and major leaps forward in sustainability. More recently, different practitioners from multiple disciplinary backgrounds have endeavoured to develop multidisciplinary and transdisciplinary approaches to create sustainability science as a means of addressing the need for transformations towards more sustainable practices. These approaches have included ecological modernization (Mol and Sonnenfield, 2000), exploration of resilience (Berkes *et al.*, 2003; Darnohofer *et al.*, 2010; Wilson, 2012) and Ostrom's (2005) Institutional Analysis and Development (IAD) framework. These approaches have in common the emphasis on transdisciplinary perspectives on resource management, the need for transformative action in response to compromised socio-ecological sustainability, the importance of institutional analysis at multiple scales and the need for stakeholder engagement in an iterative, collaborative and comprehensive way. In the 2000s, the literature on sustainable niche management, transition management and the MLP has grown to prominence, founded on science and technology studies.

Assessment of transition in the agricultural sector has been somewhat less developed, with the few studies tending to focus on the development of niche innovations (e.g. Wiskerke and van der Ploeg, 2004; Knickel *et al.*, 2009), organic farming (Belz, 2004; Smith, 2006, 2007), individual farm-level transitions (Wilson, 2007, 2008) or specific national phenomena (Poppe *et al.*, 2009). In this book, we focus on 'emerging transitions' – niches which have developed past the novelty stage and which are impacting on the agricultural sector at regional level. Our interest is on niche-regime interactions, in other words how niches and regimes co-develop in response to changed socio-technical landscape conditions. Change in agricultural systems have been widely explored at the macro level in recent literature on food regimes, particularly emphasizing the roles of markets, large retailers and government import/export strategies in the development of agri-food systems (Friedmann, 2009; McMichael, 2009). There is also an extensive body of literature at the meso level, addressing changes in farming systems, and the role of retailers, processors and consumers in the production and consumption chains. At the micro (farm) level, sustainable transitions have been identified through a variety of approaches, most directly through Wilson's (2007, 2008) analysis of farm transitions in relation to multifunctionality. Where the MLP holds particular utility is in its conceptualization of multiple societal levels, demonstrating the interaction of niche innovations, mainstream regime actors and landscape-level factors.

As Chapter 2 (Darnhofer, this volume) will outline in detail, in this book we use as a foundation the MLP (e.g. Geels *et al.*, 2004; Geels, 2005; Kemp and Martens, 2007; Kemp *et al.*, 2007; Rotmans and Loorbach, 2009; Geels and Schot, 2010), which recognizes the importance of taking into account interrelations between societal domains and complexity – that is, uncertainty and unpredictability – in understanding change processes. The principles of the MLP, although primarily developed in relation to technological transition, fit well with recent research on resilience and adaptive capacity as indicators of sustainability in agricultural systems. Key to MLP conceptions is the notion that transition involves changes in socio-technical systems: "a cluster of elements, including technology, regulations, user practices and markets, cultural meanings, infrastructure, maintenance networks and supply networks" (Geels *et al.*, 2004:3). This system is conceptualized as representing a stable 'regime' – a semi-coherent set of rules which provide orientation and coordination to actor groups around a specific function. The regime is thus 'locked in' and changes only incrementally. Radical transition is conceptualized as occurring in niches – protected spaces where novelty innovations are developed by supportive actors. These niches may exist for some time before a change in the 'socio-technical landscape' (the environment beyond the direct influence of niche and regime actors, such as macro-economics, climate, deep cultural patterns) create pressure on the mainstream system which creates a 'window of opportunity' for niche innovations to influence the regime. It is this process of niche-regime interaction that is the focus of this book.

Structure of the book

This book is based on research findings from the FarmPath (Farming Transitions: Pathways towards regional sustainability of agriculture in Europe) project, funded by the European Commission's 7th Framework Programme. The research was undertaken in response to a call to identify and assess innovative models and approaches to agriculture, with a view to identifying potential transition pathways, and options for disseminating and scaling up these approaches within the European context. The book chapters thus draw on a single conceptual framework, rooted in the MLP, which is described in detail in Chapter 2 (Darnhofer, this volume). Chapter 3 establishes the backdrop for regional-level transition in European agriculture, through an analysis of trends in the agricultural sector and rural areas over the past 30 years, proposing a typology of rural areas in Europe (Slee and Pinto-Correia, this volume). In Chapter 4, the methodological underpinnings of the case study research are described, with an emphasis on lessons learned from applying the MLP in empirical field research (Karanikolas *et al.*, this volume). To date, most analyses of transition processes utilizing the MLP have been conceptual or based on document review. Chapters 5 through 11 focus on findings from empirical case studies. In each chapter, three regional level case studies are contrasted, selected from the seven study countries: Bulgaria, the Czech Republic, France, Germany, Greece, Portugal and the United Kingdom (Fig. 1.1).

Fig. 1.1. Case study countries (Source: authors).

In Chapter 5, the emergence of 'lifestyle farming' in response to countryside consumption in peripheral rural areas is explored, demonstrating the development of an 'unseen' niche which has emerged without the overt protection of supportive actors (Pinto-Correia *et al.*a, this volume). Chapter 6 provides insights into the role of collaboration in transition processes (Schiller *et al.*, this volume). Chapter 7 explores the emergence of 'high nature value' farming, contrasting processes of niche anchoring between 'top-down' and 'bottom-up' initiatives (Peneva *et al.*, this volume). Transitions relating to public goods from the agricultural sector, particularly water quality, are addressed in Chapter 8 (Vlahos and Schiller, this volume). In Chapter 9, analysis of on-farm renewable energy production demonstrates how technologies developed in the agricultural sector have been 'translated' in response to changing landscape pressures on the energy regime (Sutherland *et al.*b, this volume). The assessment of alternative agri-food networks in Chapter 10 demonstrates the varying sources of innovation, and forms of resistance adopted by regime actors to niche innovations (Darrot *et al.*, this volume). In Chapter 11, local certification schemes are assessed in relation to the role that new forms of governance can play in facilitating sustainability transitions (Lošťák *et al.*, this volume). Chapter 12 considers the role of transdisciplinarity in establishing future pathways for regional transition, describing the development of a process for identifying 'transition pathways' (Pinto-Correia *et al.*b, this volume). We conclude the book with a synthesis of conceptual lessons learned (Darnofer *et al.*, Chapter 13, this volume), and of empirical findings and challenges for future research (Sutherland *et al.*c, Chapter 14, this volume).

References

Belz, F.M. (2004) A transition towards sustainability in the Swiss agri-food chain (1970-2000): Using and improving the multi-level perspective. In: Elzen, B., Geels, F.W. and Green, K. *System Innovation and the Transition to Sustainability.* Edward Elgar, Cheltenham, UK, pp. 97–113.

Berkes, F., Colding, J. and Folke, C. (eds) (2003) *Navigating Socio-Ecological Systems: Building Resilience for Complexity and Change.* Cambridge University Press, Cambridge, UK.

Boström, M. (2012) A missing pillar? Challenges in theorizing and practicing social sustainability: Introduction to the special issue. *Sustainability: Science, Practice, & Policy* 8, 3–14.

Brouwer, F. (ed.) (2004) *Sustaining Agriculture and the Rural Environment: Governance, Policy and Multifunctionality*. Edward Elgar, Cheltenham, UK.

Brouwer, F. and van der Heide, C.M. (2009) *Multifunctional Rural Land Management: Economics and Policies*. Earthscan, London, UK.

Buller, H., Wilson, G.A. and Höll, A. (eds) (2000) *Agri-Environmental Policy in the European Union*. Ashgate, Aldershot, UK.

Campbell, H. (2005) The rise and rise of Eurep-GAP: European (re)invention of colonial food relations? *International Journal of Sociology of Agriculture and Food* 13, 1–19.

Clark, J. (2005) The 'New Associationalism' in agriculture: Agro-food diversification and multifunctional production logics. *Journal of Economic Geography* 5, 475–498.

Clark, J. (2006) The institutional limits to multifunctional agriculture: Subnational governance and regional systems of innovation. *Environment and Planning C: Government and Policy* 24, 331–349.

Cline, W. (2007) *Global Warming and Agriculture: Impact Estimates by Country*. Peterson Institute, Washington DC.

Council Regulation (EC) No 1257/1999. *Official Journal of the European Communities* 26.2.1999. Available from http://www.esf.ie/downloads/

Darnhofer, I. (2015) Socio-technical transitions in farming. Key concepts. In: Sutherland, L-A., Darnhofer, I., Wilson, G.A. and Zagata, L. (eds) *Transition Pathways towards Sustainability in Agriculture: Case Studies from Europe*. CABI, Wallingford, UK, pp. 17–32.

Darnhofer, I., Bellon, S., Dedieu, B. and Milestad, R. (2010) Adaptiveness to enhance the sustainability of farming systems. A review. *Agronomy for Sustainable Development* 30, 545–555.

Darnhofer, I., Sutherland, L-A. and Pinto-Correia, T. (2015) Conceptual insights derived from case studies on 'emerging transitions' in farming. In: Sutherland, L-A., Darnhofer, I., Wilson, G.A. and Zagata, L. (eds) *Transition Pathways towards Sustainability in Agriculture: Case Studies from Europe*. CABI, Wallingford, UK, pp. 189–204.

Darrot, C., Diaz, M., Tsakalou, E. and Zagata L. (2015) 'The Missing Actor': Alternative agri-food networks and the resistance of key regime actors. In: Sutherland, L-A., Darnhofer, I., Wilson, G.A. and Zagata, L. (eds) *Transition Pathways towards Sustainability in Agriculture: Case Studies from Europe*. CABI, Wallingford, UK, pp. 143–156.

DGIP (2012) EU *Measures to Encourage and Support New Entrants.* Note published by the Directorate-General for Internal Policies. Available from: http://www.europarl.europa.eu/committees/en/agri/

Dibden, J. and Cocklin, C. (2009) 'Multifunctionality': Trade protectionism or a new way forward? *Environment and Planning A* 41, 163–182.

European Commission (2009) *Adapting to Climate Change: Towards a European Framework for Action*. Available from: http://eur-lex.europa.eu/legal-content/EN/TXT/PDF/

European Commission (2012) *Europe's Path Towards Sustainable Agriculture*. Speech 12/480 delivered by Dacian Cioloş at the 'Agriculture the way towards sustainability and inclusiveness' G20/Rio de Janeiro 21 June 2012.

European Environment Agency (2006a) *The Changing Faces of Europe's Coastal Areas*. EEA Report No 6/2006. OPOCE (Office for Official Publications of the European Communities).

European Environment Agency (2006b) *Urban Sprawl in Europe. The Ignored Challenge*. EEA Report No 10/2006. OPOCE (Office for Official Publications of the European Communities).

European Environment Agency (2006c) *Land Accounts for Europe* 1990-2000, EEA Report No 11/2006. OPOCE (Office for Official Publications of the European Communities).

Eurostat (2011) Agricultural Census 2010 – Provisional Results. Available from: http://epp.eurostat.ec.europa.eu/statisticsexplained/index.php/Agriculturalcensus_2010_-_provisionalresults

Friedmann, H. (2009) Moving food regimes forward. *Agriculture and Human Values* 26, 335–344.

Geels, F.W. (2005) Processes and patterns in transition and system innovations: Refining the co-evolutionary multi-level perspective. *Technological Forecasting and Social Change* 72, 681–696.

Geels, F.W. and Schot, J. (2010) The dynamics of transitions. A socio-technical perspective. In: Grin, J., Rotmans, J. and Schot, J. *Transitions to Sustainable Development*. Routledge, New York, pp. 11–101.

Geels, F.W., Elzen, B. and Green, K. (2004) General introduction: System innovation and transitions to sustainability. In: Elzen, B., Geels, F.W. and Green, K. *System Innovation and the Transition to Sustainability*. Edward Elgar, Cheltenham, UK, pp. 1–16.

Gray, J. (2002) *False Dawn: The Delusions of Global Capitalism,* 2nd edn. Granta Books, London, UK.

Heinberg, R. and Bomford, D. (2009) *The Food and Farming Transition. Towards a Post Carbon Food System*. Post Carbon Institute, Santa Rosa, California.

Imeson, A. (2012) *Desertification, Land Degradation and Sustainability: Paradigms, Processes, Principles and Policies*. Wiley-Blackwell, Oxford, UK.

Karanikolas, P., Vlahos, G. and Sutherland, L-A. (2015) Utilizing the multi-level perspective in empirical field research: Methodological considerations. In: Sutherland, L-A., Darnhofer, I., Wilson, G.A. and Zagata, L. (eds) *Transition Pathways towards Sustainability in Agriculture: Case Studies from Europe*. CABI, Wallingford, UK, pp. 51–66.

Kemp, R. and Martens, P. (2007) Sustainable development: How to manage something that is subjective and never can be achieved? *Sustainability: Science, Practice & Policy* 3, 5–14.

Kemp, R., Loorbach, D. and Rotmans, J. (2007) Transition management as a model for managing processes of co-evolution towards sustainable development. *International Journal of Sustainable Development and World Ecology* 14, 78–91.

Knickel, K., Tisenkopfs, T. and Peter, S. (eds) (2009) *Innovation Processes in Agriculture and Rural Development. Results of a Cross-national Analysis of the Situation in Seven Countries, Research Gaps and Recommendations*. INSIGHT FP6 Project Final Report – Comparative Analysis and Synthesis.

Lang, T. and Heasman, M. (2004) *Food Wars: The Global Battle for Mouths, Minds and Markets*. Earthscan, London, UK.

Loorbach, D. and Frantzeskaki, N. (2009) *A Transition Research Perspective on Governance for Sustainability*. Paper presented at the conference: Sustainable development: A challenge for European research. 28-29 May 2009, Brussels, Belgium.

Lošťák, M., Karanikolas, P., Draganova, M. and Zagata, L. (2015) Local quality and certification schemes as new forms of governance in sustainability transitions. In: Sutherland, L-A., Darnhofer, I., Wilson, G.A. and Zagata, L. (eds) *Transition Pathways towards Sustainability in Agriculture: Case Studies from Europe*. CABI, Wallingford, UK, pp. 157–170.

Marsden, T. (2003) *The Condition of Rural Sustainability*. Van Gorcum, Assen, the Netherlands.

Marsden, T. (2013) From post-productionism to reflexive governance: Contested transitions in securing more sustainable food futures. *Journal of Rural Studies* 29, 123–134.

Marsden, T., Murdoch, J., Lowe, P., Munton, R. and Flynn, A. (1993) *Constructing the Countryside*. UCL Press, London, UK.
McMichael, P. (2009) A food regime analysis of the 'world food crisis'. *Agricultural and Human Values* 26, 281–295.
Mestre-Sanchis, F. and Feijoo-Bello, M.L. (2009) Climate change and its marginalizing effect on agriculture. *Ecological Economics* 68, 896–904.
Mol, A.P.J. and Sonnenfield, D.A. (2000) *Ecological Modernisation Around the World: Perspectives and Critical Debates*. Frank Cass, London, UK.
Murphy, K. (2012) The social pillar of sustainable development: A literature review and framework for policy analysis. *Sustainability: Science, Practice, & Policy* 8, 15–29.
Ostrom, E. (2005) *Understanding Institutional Diversity*. Princeton University Press, Princeton.
Peneva, M., Draganova, M., Gonzalez, C., Diaz, M. and Mishev, P. (2015) High nature value farming: Environmental practices for rural sustainability. In: Sutherland, L-A., Darnhofer, I., Wilson, G.A. and Zagata, L. (eds) *Transition Pathways towards Sustainability in Agriculture: Case Studies from Europe*. CABI, Wallingford, UK, pp. 97–112.
Pinto-Correia, T., Gonzalez, C., Sutherland, L-A. and Peneva, M. (2015a) Lifestyle farming: Countryside consumption and transition towards new farming models. In: Sutherland, L-A., Darnhofer, I., Wilson, G.A. and Zagata, L. (eds) *Transition Pathways towards Sustainability in Agriculture: Case Studies from Europe*. CABI, Wallingford, UK, pp. 67–82.
Pinto-Correia, T., McKee, A. and Guimarães, H. (2015b) Transdisciplinarity in deriving transition pathways for agriculture. In: Sutherland, L-A., Darnhofer, I., Wilson, G.A. and Zagata, L. (eds) *Transition Pathways towards Sustainability in Agriculture: Case Studies from Europe*. CABI, Wallingford, UK, pp. 171–188.
Poppe, K.R., Termeer, C. and Slingerland, M. (eds) (2009) *Transitions towards Sustainable Agriculture and Food Chains in Peri-Urban Areas*. Wageningen Academic Publishers, Wageningen, the Netherlands.
Potter, C. and Burney, J. (2002) Agricultural multifunctionality in the WTO: Legitimate non-trade concern or disguised protectionism? *Journal of Rural Studies* 18, 35–47.
Potter, C.A. and Tilzey, M. (2005) Agricultural policy discourses in the European post-Fordist transition: Neoliberalism, neomercantilism and multifunctionality. *Progress in Human Geography* 29, 1–20.
Potter, C. and Tilzey, M. (2007) Agricultural multifunctionality, environmental sustainability and the WTO: Resistance or accommodation to the neoliberal project for agriculture? *Geoforum* 38, 1290–1303.
Pretty, J.N. (2002) *Agri-culture: Reconnecting People, Land and Nature*. Earthscan, London, UK.
Robinson, G. (ed.) (2008) *Sustainable Rural Systems: Sustainable Agriculture and Rural Communities*. Ashgate, Aldershot, UK.
Rotmans, J. and Loorbach, D. (2009) Complexity and transition management. *Journal of Industrial Ecology* 13, 184–196.
Slee, B. and Pinto-Correia, T. (2015) Understanding the diversity of European rural areas. In: Sutherland, L-A., Darnhofer, I., Wilson, G.A. and Zagata, L. (eds) *Transition Pathways towards Sustainability in Agriculture: Case Studies from Europe*. CABI, Wallingford, UK, pp. 33–50.
Smith, A. (2006) Green niches in sustainable development: The case of organic food in the United Kingdom. *Environment and Planning C: Government and Policy* 24, 439–458.
Smith, A. (2007) Translating sustainabilities between green niches and socio-technical regimes. *Technology Analysis & Strategic Management* 19, 427–450.
Stiglitz, J. (2002) *Globalization and its Discontents*. Penguin, London, UK.
Sutherland, L-A., Peter, S. and Zagata, L. (2015b) On-farm renewable energy: A 'classic case' of technological transition. In: Sutherland, L-A., Darnhofer, I., Wilson, G.A. and Zagata, L. (eds) *Transition Pathways towards Sustainability in Agriculture: Case Studies from Europe*. CABI, Wallingford, UK, pp. 127–142.
Sutherland, L-A., Zagata, L. and Wilson, G.A. (2015c) Conclusions. In: Sutherland, L-A., Darnhofer, I., Wilson, G.A. and Zagata, L. (eds) *Transition Pathways towards Sustainability in Agriculture: Case Studies from Europe*. CABI, Wallingford, UK, pp. 205–220.

van der Ploeg, J.D. (2007) *Empire and the Peasant Principle.* Paper presented at the plenary session of the XXI Congress of the European Society for Rural Sociology. 22-26 August 2007, Keszthely, Hungary.

van Huylenbroek, G. and Durand, G. (eds) (2003) *Multifunctional Agriculture: A New Paradigm for European Agriculture and Rural Development.* Ashgate, Aldershot, UK.

Vlahos, G. and Schiller, S. (2015) Transition processes and natural resource management. In: Sutherland, L-A., Darnhofer, I., Wilson, G.A. and Zagata, L. (eds) *Transition Pathways towards Sustainability in Agriculture: Case Studies from Europe.* CABI, Wallingford, UK, pp. 113–126.

Wilson, G.A. (2001) From productivism to post-productivism ... and back again? Exploring the (un)changed natural and mental landscapes of European agriculture. *Transactions of the Institute of British Geographers* 26, 77–102.

Wilson, G.A. (2007) *Multifunctional Agriculture. A Transition Theory Perspective.* CABI, Wallingford, UK.

Wilson, G.A. (2008) From 'weak' to 'strong' multifunctionality: Conceptualising multifunctional transitional pathways. *Journal of Rural Studies* 24, 367–383.

Wilson, G.A. (2010) Multifunctional 'quality' and rural community resilience. *Transactions of the Institute of British Geographers* 35, 364–381.

Wilson, G.A. (2012) *Community Resilience and Environmental Transitions.* Routledge, London, UK.

Wiskerke, J.S.C. and van der Ploeg, J.D. (eds) (2004) *Seeds of Transition. Essays on Novelty Production, Niches and Regimes in Agriculture.* Van Gorcum, Assen, the Netherlands.

Woods, M. (2005) *Rural Geography: Processes, Responses and Experiences in Rural Restructuring.* Sage, London, UK.

Zagata, L. and Lošťák, M. (2014) *Farming Models for Young Farmers and New Entrants.* Final WP4 Report. Final Report for the FarmPath 7th Framework Research Project. Available at: www.farmpath.eu

Chapter 2

Socio-technical transitions in farming: key concepts

I. Darnhofer[1]

[1]*University of Natural Resources and Life Sciences, Vienna (ika.darnhofer@boku.ac.at)*

Introduction

Transition studies usually focus on processes of radical change at the level of a country, or even at the international level. Since a transition can only be ascertained in hindsight, many of these studies are historical. The case studies presented in this book differ from many other studies on transition to sustainability in several ways: the focus is on emerging transitions (transitions in-the-making) and we focus on a smaller spatial level, that of the region. These specifications allow us to zoom-in on the niche-regime interactions that play a key role in the 'take-off' phase of a transition; when a niche engages with the regime to initiate radical technical, institutional and structural changes.

The case studies presented in this book were selected because they have the potential to contribute to a transition to sustainability in agriculture. They focus on new developments that question the dominant paradigm, calling into question the basic assumptions of the existing regime in a fundamental way. To achieve radical change, the niche is involved in a whole complex of interrelated changes including, for example, new beliefs and values; new technologies and practices; new configurations of actor groups; new networks or new policies.

This chapter presents the core characteristics of studies of 'socio-technical transitions to sustainability', focusing on the multi-level perspective (MLP) and niche-regime linking. It includes a very brief overview of previous work on niche-internal processes, which was situated within research on (endogenous) rural development. Finally, some core concepts on transitions within farming will be discussed: the need of a niche to cope with the complexity and unpredictability of societal change; the role of power in resisting and steering transitions; the politics involved in defining sustainability in agriculture; and the role of changing rules and values to engender transitions.

Socio-technical transitions to sustainability

Characteristics of transition studies

Transition studies build on a wide range of theoretical backgrounds (Geels and Schot, 2010). These include evolutionary economics, which focuses on long-term processes and developed the concept of technological regime to understand coordination within a population of firms. They also include sociology, especially structuration theory (Giddens, 1984), which assumes knowledgeable, interpretive actors that enact rules and structures, and where structures guide but do not determine action. Furthermore, they draw heavily on innovation studies and on science and technology studies (STS) which have shown the complexity, fluidity and contingency of technological change (Elzen *et al.*, 2004b).

The concept of co-evolution denotes the interaction between societal subsystems which influence the dynamics of the societal system under study. Indeed, as economic, cultural, technological, ecological and institutional subsystems interact, they respond to changes in each other and adapt. Understanding transitions thus means that structures, cultures and practices of a societal system are analysed in an integrative manner (de Haan and Rotmans, 2011). Structures include the formal, physical, legal and economic aspects that enable or restrict practices. Cultures include the cognitive, discursive and ideological aspects involved in sense-making. Finally, practices include the routines, habits and procedures through which actors (individuals, organizations) maintain the functioning of the societal system. Since structures, cultures and practices co-evolve, it implies that in a transition, they are fundamentally changed so that the way the societal system functions is profoundly altered (de Haan and Rotmans, 2011).

Within studies of socio-technical transitions, two broad approaches can be distinguished. First, there are historical studies of completed socio-technical transitions (e.g. the shift from sailboats to steamships, Geels, 2002 or from horse-drawn carriages to automobiles, Geels, 2005). They were driven by the commercial motivation of pioneers and entrepreneurs who developed the technology. They were not planned or managed by policy. Their objectives were not determined beforehand, but the transitions and their directions emerged as a result of co-evolutionary processes involving a variety of societal influences (Slingerland and Rabbinge, 2009). Thus, while normative changes were often involved, they were not the main drivers.

Second, there are studies of *current societal changes*. These often explicitly focus on 'transitions to sustainability', which is a normative goal and thus there is an (implicit) intention to steer them in the 'right' direction (Grin *et al.*, 2010). As these are ongoing processes and future developments cannot be predicted, it is uncertain whether the outcome will be limited to incremental change or whether there will be a radical transformation. Whilst for historical studies it is clear what changes were brought about, for studies of ongoing processes it remains a challenge to distinguish incremental change processes which are ongoing in any regime, from those change processes which will – eventually – lead to systemic, radical change and thus qualify as 'transition'.

The multi-level perspective

The MLP views transitions as non-linear processes that result from the interplay of developments at three analytical levels: niches (the locus of radical innovations);

socio-technical regimes (the locus of established practices and associated rules that stabilize existing systems); and an exogenous socio-technical landscape (Geels, 2011). Each level refers to a heterogeneous configuration of elements, with the regime more stable than the niche in terms of number of actors and degrees of alignment between the elements (Geels, 2011). The MLP emphasizes that for a transition to be successful, processes at the niche, regime and landscape level need to be aligned.

The three levels might at times seem like spatial concepts, not least because a regime is often studied at the national level and socio-technical landscapes often refer to international trends. However, the three levels are actually defined by their (relative) temporal stability, not by their spatial spread. Nonetheless in practice, the two dimensions are often related as practices that involve a wide variety of societal actors tend to be spread over larger areas, and tend to be stable over time. On the other hand, smaller networks may be more dependent on individual actors or susceptible to shocks and thus are less stable.

Niches are created by actors at the local level, for example through the invention of a new technology, or entrepreneurs developing a new market. They may be protected spaces, such as subsidized demonstration projects, or small market niches where users have special demands and are willing to support emerging innovations (such as local organic food chains). Niche innovations are often characterized by a mismatch with the existing regime, lack of appropriate infrastructure, regulations, or incompatibility with consumer routines. Over time some niches stabilize through activities such as: the articulation and adjustment of expectations or visions; building of social networks and the enrolment of more actors; as well as learning processes on issues such as technical design, user preferences, organizational issues and business models, policy instruments and symbolic meanings (Schot and Geels, 2008). Niches are crucial for transitions because they provide the seeds for systemic change, even if many of these seedlings will eventually perish (Elzen *et al.*, 2004a).

The regime is the meso-level and is of central importance for transition research since it defines the societal systems within which transitions are mainly analysed. The regime includes both the tangible and measurable elements (such as artefacts, market shares, infrastructure, regulations, consumption patterns, public opinion) as well as intangible elements. This includes the deep structure made up of beliefs, rules of thumb, routines, and standardized ways of doing things, policy paradigms, social expectations and norms (Geels, 2011). A regime is characterized by fairly stable rules such as cognitive routines, shared beliefs, capabilities and competencies, lifestyles and user practices, institutional arrangements and regulations, and legally binding contracts. Since these elements, as well as items such as physical infrastructures and organizations, are well aligned, regimes are characterised by lock-in. Innovation occurs incrementally with small adjustments accumulating into stable trajectories. A regime is composed of several sub-regimes (such as user preferences, market, policy, and science) which have their own dynamics but interpenetrate and co-evolve with each other. The concept of a socio-technical regime aims to capture the meta-coordination between these different sub-regimes (Geels, 2004).

The socio-technical landscape designates the long-term exogenous trends at the macro-level. This includes demographic trends, political ideologies, societal values, macro-economic patterns, climate change. In the short-term, these processes at the level of the socio-technical landscape cannot be influenced by niche or regime actors (Geels and Schot, 2010).

Since the existing socio-technical regime is stabilized in many ways (for example through technical standards, sunk costs by key players who have no incentive to change, production structures, industry networks, and user practices), transitions do not come about easily. However, over time a regime will display weaknesses, often as a result of unintended side-effects that accumulate and become problematic. As persistent problems become increasingly obvious, they can lead to pressure from the socio-technical landscape to alter practices. This creates a window of opportunity for a niche to break through, especially if the regime is not able to adequately address these persistent problems.

When analysing emerging transitions it is particularly important to keep in mind that regimes are not homogeneous and/or monolithic. Indeed, whereas regimes may appear as coherent blocks from the outside (and tend to represent themselves that way), there are often internal tensions, disagreement and conflicts of interest (Geels, 2011). An example may be the organizations that represent agriculture (e.g. Ministry of Agriculture, Chamber of Agriculture) who tend to project a unified image. Yet, not all farmers adhere to the modernization paradigm and a number follow different logics and farming styles. Thus, regimes have coherence, shared values and similarity but on the other hand contain variety, disagreement on specific issues, debate and internal conflict (Geels, 2011). These tensions can be an opportunity for niche actors to identify sympathetic regime actors and gain support (Diaz *et al.*, 2013).

Furthermore, when analysing emerging transitions it is important to pay attention to the interactions between regimes (Geels, 2011). Indeed, the growth of niches often requires interaction between two (or more) regimes, for example between agriculture and energy regimes in the case of biofuels. Thus, the positive or negative interaction between regimes can enable an emerging transition. However, niche actors behind an emerging transition may also actively construct new relationships between previously separate regimes, as part of their proposal to radically alter the dominant regime.

Anchoring: initiating and strengthening links between a niche and the regime

Smith (2007) has pointed out that there has been insufficient consideration of the processes by which niches and regimes interact and are interdependent, and has therefore focused on the 'linking' between niche and regime. Elzen *et al.* (2012) proposed the term 'anchoring' for emerging forms of linking, reserving the term 'linking' for later stages when the interaction is more robust.

Building on the three analytical dimensions that can be distinguished when analysing the regime (see Fig. 2.1) Elzen *et al.* (2012) propose that anchoring can occur in each of these dimensions. They therefore suggest distinguishing between 'technological anchoring', such as when the technological characteristics are further defined; 'network anchoring' where changes occur in the network of actors that carry the niche (for instance through expansion of the network or intensified exchanges); and 'institutional anchoring' where new rules are developed, including new beliefs, new views of a problem, or new regulations, standards or policies. Understanding the variety of ways in which a niche may anchor and then link to a regime (or several regimes) is relevant for better understanding how desirable changes may be induced (Elzen *et al.*, 2012).

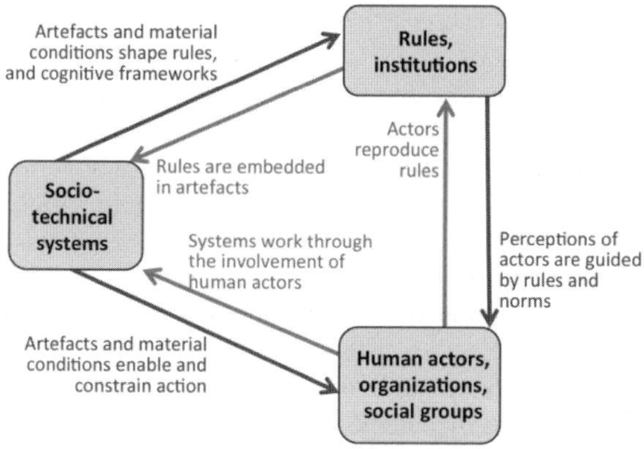

Fig. 2.1 Three analytic dimensions to understand regimes and identify potential 'anchors' to link niches and regimes and thus induce transitions (Source: adapted from Geels, 2004:903, reprinted with permission).

Studies of transitions in agriculture

Research on processes at niche-level: novelties

There is a vast literature on rural development in Europe, closely linked to multifunctionality and thus to social and economic sustainability of farming. Most of this literature focuses on processes at the level of novelties, such as on-farm processing and direct marketing. Such farm level processes were extensively studied in the framework of (endogenous) rural development (e.g. van der Ploeg and Marsden, 2008). In this context, the term 'transition' has been used to indicate a reconfiguration of activities at the farm level (Wilson, 2007; Lamine and Bellon, 2009; Milone, 2009) but does not indicate changes in dominant practices at regional or national levels.

This work on (endogenous) rural development generally does not build on the conceptual framework of transition studies. However, many rural development initiatives are the site where novelties are developed (Roep and Wiskerke, 2004; Schmid *et al.*, 2004; Oostindie and van Broekhuizen, 2008) and if the novelty stabilizes it may lead to the establishment of a niche, as conceptualized in the framework of transition studies. Most of these niches, even if they are a form of resistance or contestation of the dominant regime, have not (yet) significantly changed it, much less induced a transition. Thus, while such niches play a crucial role in developing alternatives and demonstrating that they propose a viable system, it is not clear under which conditions they 'break through' and induce a transition (assuming they aim to do so). This, however, is the core issue in research on transitions towards sustainability: the dynamics that fundamentally alter dominant practices, replacing the incumbent regime by realigning technical processes, social actors and mental frameworks.

Studies using the MLP

There have also been a few studies that have analysed historical transitions in agriculture from a MLP perspective. Some are based on available historical literature and study long-term transitions (50 years or more). For example Grin (2010) has analysed the modernization of agriculture in the Netherlands between 1886 and 2006 (see also Geels, 2009). Belz (2004) has retraced the transition in Switzerland from intensive, industrialized practices towards integrated and organic practices; Smith (2006) has analysed the development of organic farming in the UK; and Sinclair (2014) has retraced the transformative change in the Australian subtropical dairy system, following its deregulation. Other work done on agriculture and agri-food chains focuses mostly on niche-level process, such as how novelty emerges and how a niche stabilizes. Some of this work has been published in edited books (Elzen *et al.*, 2004b; Wiskerke and van der Ploeg, 2004; Poppe *et al.*, 2009; Vellema, 2011; Spaargaren *et al.*, 2012). In summarizing the factors that encourage the stabilization of niches in agriculture, Roep and Wiskerke (2004) identify seven lessons:

- Create and maintain a learning environment, privileging double-loop learning (learning about the assumptions, meaning and preferences of relevant actors). Learning should focus on how networks are built and maintained, and on the complex interactions between technical and institutional aspects linked to novelty creation;
- Explore and understand diversity, which enables niche actors to identify and present a novelty as promising; and to develop it into a convincing and well-functioning programme;
- Make new and effective connections, for example with rural entrepreneurs, researchers, extension agents, or farmers' unions;
- Take into account that creating alignment between strategies and expectations is a continuous process, thus the niche and its network requires continuous management and evaluation aimed at maintaining individual responsibility and commitment to the collective goal;
- Ensure that all actors improve their own situation, since progress or reciprocity (at the material or the moral economy level) is the reason for their participation;
- Recognize change agents as crucial to set the process in motion, through actions such as envisioning windows of opportunity, expressing expectations, enrolling alliances and creating room for manoeuvre at the local level; and
- Appreciate the value of the unexpected. In other words building the capacity to transform the unexpected or unintended into something useful or valuable, instead of assessing outcomes only according to initial expectations and learning processes.

Similar conclusions were reached based on the study of a wide range of organic marketing initiatives (Schmid *et al.*, 2004), where the internal factors for success were found to include: professional management, key individuals, clear objectives and strategic planning, recognition of strategic turning points, ensuring motivation and coherence, innovation and market research, and networking. Tisenkopfs *et al.* (2009) reached similar conclusions based on their analysis of how novelties mature into a niche (how innovation networks are constructed). They highlighted the relevance of activities such as networking;

multi-actor participation; individual and collective learning; building on the interplay between economic, technical, organizational and social innovations; as well as the importance of collaboration and territorial governance (Tisenkopfs *et al.*, 2009). A longitudinal study of the development of a niche showed how it was influenced by the evolution of its network (Hermans *et al.*, 2013), an evolution that was driven by internal processes of convergence of expectations, as well as the learning and experimentation processes resulting from the projects it undertook.

Other studies (e.g. Bos and Grin, 2008; Elzen *et al.*, 2011) have focused on pressures from the socio-technical landscape and on niche-regime interactions that have enabled (or not) the establishment of a niche such as, for example, pressure on pig husbandry systems to comply with animal welfare and sustainability concerns. These studies can be seen as first case studies of ongoing ('emerging') transitions. They highlight how processes at each of the three analytical levels need to align for an innovation to 'break through' and transform the regime.

FarmPath: emergent regional-level transitions to sustainability

Overlapping niches and regimes

FarmPath focuses on emerging transitions. In the framework of the research project, these are defined as established niches that have engaged with actors and organizational structures at the regime level in a significant way. That a niche is 'established' implies that there is a stable pattern (rules and standards are defined, networks are established, etc.). To initiate a 'transition to sustainability' an established niche would need to seek radical change. This is change that:

- affects a whole sector, a whole value chain, or a whole territory;
- leads to a new alignment of actors, networks or regimes;
- is based on rules and values that are clearly distinct from those of the regime;
- addresses a sustainability issue that is clearly defined by the actors involved in the emerging transition.

Studies building on the MLP have tended to emphasize the role of structure (e.g. the role of technological innovations), while at times underplaying the role of 'soft factors' such as consumer preferences, beliefs and power structures (Elzen *et al.*, 2004a; Holtz *et al.*, 2008; Geels and Schot, 2010). However, transitions to sustainability in agriculture may not be primarily technology-driven. It is likely that social innovation will play an important role, and that radical change in agriculture will require changes in beliefs and values by a wide range of societal actors. The focus of the analysis in this book is on capturing these social processes, especially in the interaction between the niche and the regime.

Conceptually, to analyse niche-regime interactions the analysis builds on the notion of 'anchoring' (Elzen *et al.*, 2012) and 'linking' (Smith, 2007) (Fig. 2.2). Focusing on how the niche and the regime are interdependent and interact, these authors propose a less hierarchical representation of niches and regimes. By considering that they may overlap (Fig. 2.2., see also Diaz *et al.*, 2013), it allows for 'hybrid actors' (regime actors that are sympathetic to the proposals of the niche, and are symptomatic of the heterogeneity and dissent within a regime).

The territorial focus in FarmPath has been selected for two reasons. First, farming is necessarily spatially embedded and thus affected by the nature of those spatial conditions. These include natural conditions (soils, agri-ecosystem, climate etc.) as well as social and cultural networks, values and traditions. As such, what constitutes sustainable practices will necessarily be site-specific and cannot be defined in a 'one size fits all' approach. Second, the focus on regional-level transitions allows us to 'zoom in' on the processes of 'anchoring' and 'linking', processes through which the niche actors interact with the regime, and may initiate radical changes. The analytical focus is therefore on these processes: how niche actors take advantage of complexity and unpredictable change; the power issues involved in achieving change; the (re-)negotiation of the definition of sustainability; and the strategies to bring about changes in institutions.

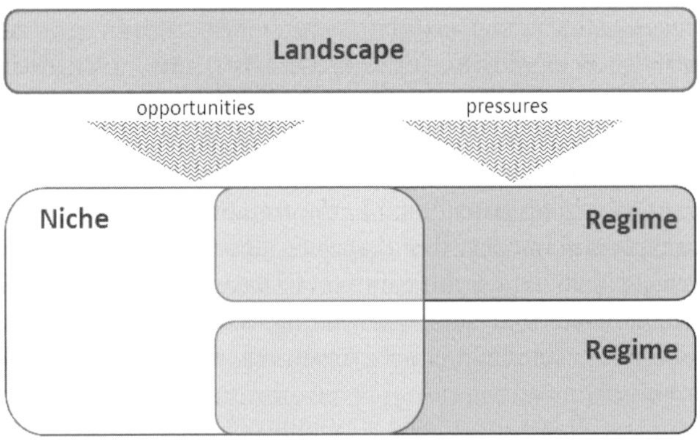

Fig. 2.2. The landscape puts pressure on the regime(s) to address a persistent problem, and this pressure can present opportunities for the niche to 'take off'. To represent the 'anchoring' and 'linking' activities of the niche, it is helpful to conceptualize an area of overlap between the regime and a niche. This is where niche actors link to 'hybrid actors'. In this figure, the innovation by the niche also links two previously separate regimes (e.g. agriculture and energy). (Source: authors).

Complexity: riding the dynamics

Transitions may sometimes be seen in a mechanistic way, as a set of factors or conditions that – if they all work together – will cause a desired change. The underlying implication is that transition processes can be steered or engineered to a certain extent. Thus, even if guidelines derived from transition studies are not deterministic, recommendations, methods, and techniques are still often presumed to have real effects which can be used to attain certain objectives and solve certain problems. This, in effect, is a form of social engineering (Duineveld *et al.*, 2009).

However, there will always be differences between the situation that is analyzed and the situations for which recommendations are drawn. There will be different actors, each of whom will argue for a specific problem definition, favour different means to address the problem or prefer some outcomes over others. With every step taken by niche actors, regime actors will counter, resulting in a changed situation which demands context-specific actions by niche actors, and strategies adjusted to the revised understanding of the situation. Given the context-dependence of opportunities for action and the co-evolution of strategies, clear cause-effect relationships for inducing a transition are bound to remain elusive. As Rip (2006) has pointed out, gathering data about cause-effect relations to design an intervention with a predicted outcome contributes to the 'illusion of agency'. Indeed, based on the insights of complexity and of co-evolutionary processes driving societal change, it becomes clear that transitions cannot be technologically driven, expert-led or 'rationally' planned (Woodhill, 2009). The development of societal systems remains inherently unpredictable.

This conceptualization of complexity implies that better understanding processes will not necessarily enhance the capacity to manage. Actions (by niche or regime actors) can help steer processes in the desired direction but they do not do that by definition (Duineveld *et al.*, 2009). This implies a modest approach to the ability to 'manage' or 'steer' long-term changes in society. In other words, much caution needs to be used with the implicit assumptions conveyed by terms such as transition 'management', actors 'shaping' niches or 'selecting' one pathway over another (Shove and Walker, 2007). Many of these terms seem to imply that deliberate intervention in pursuit of specific goals is possible and potentially effective. However, care must be taken not to slip into an engineering mindset, or a belief that social change can somehow be planned and executed in a linear fashion. A transition is a long-term process involving a multitude of societal agents and is thus fraught with scientific uncertainty, social ambiguity and unpredictability.

One way forward can be to enhance learning capacities, thus enabling a greater responsiveness. In essence this means "tackling transition processes by distributing understanding, improving feedback linkages and enhancing capacities for adapting to change in a dispersed and non-hierarchical, yet coordinated, manner" (Woodhill, 2009:281). This requires capacities to design, facilitate and support such processes in ways that lead to real learning and change. For such learning networks to be successful, some critical factors can be identified, such as creating heterogeneous groups of stakeholders, developing mutual trust and social cohesion, or finding a communal perspective for the future and good process management (Vogelezang *et al.*, 2009).

The analysis of case studies can contribute to better understanding how actors have navigated complexity and unpredictability, how they have used ambiguity to their advantage, how they have been able to seize favourable moments to 'anchor' with the regime, and how they have co-constructed 'windows of opportunities'.

Power: resisting and steering transitions

A defining property of a regime is the interdependent, highly institutionalized alignment across heterogeneous processes that serves to reproduce the regime, and which tends to engender path-dependent development (Stirling and Smith, 2008). This constitutes a form of structural power which privileges certain actors at the expense of others. Indeed, some regime members command key positions in the reproduction of incumbent regimes by ensuring the maintenance of the rules, infrastructures and values underpinning socio-technical practices. However, by definition a transition to sustainability implies a radical shift. A transition leads to new technologies, social practices, institutional forms, and/or policies becoming valued. Transitions necessarily involve disrupting established personal, economic and decisional power dynamics that regime actors are likely to resist since they tend to perceive them as a loss (Kemp and van Lente, 2011). Transition processes thus need to be seen as power and relationship transformations (Duineveld *et al.*, 2009). Transitions are the result of political processes, and are ultimately legitimized and enforced through the institutions of the state. Power shifts may be a particular challenge in agriculture, a sector which is characterized by a high level of governmentality, so that transitions will be shaped by the sanctions, regulations, and styles of governance in that territory (Marsden, 2013).

Different societal players are involved in contesting and influencing the definition of what issues are seen as problematic and need to be addressed, and how they should be addressed. This gives rise to competing models for ordering the future and, depending on the path taken, different groups will win or lose (Fouilleux, 2000). As such, transitions involve social struggles such as competition, changing coalitions, and contrasting aspirations for the spatial, temporal and social distribution of benefits and costs (van der Ploeg, 2009).

Furthermore, organizations which have vested interests in the current regime may well coalesce to block policy reform that changes existing institutional and production patterns (Barbier, 2011). Many of these formal organizations in the agricultural sector tend to have vested interests in the productivist-modernization approach to agriculture and may be unwilling, or unable, to assess the relative merits of alternative paradigms (Vanloqueren and Baret, 2009; Levidow, 2011). Formal organizations are part of the dominant regime and, thus, either ignore or actively suppress the emergence of niches that may lead to new regimes in which they might lose their influence on how issues are framed, and on the approaches that are considered as efficient options for dealing with problems. Even if the process includes multi-stakeholder involvement and participative designs, these are never 'neutral' and never devoid of power and strategic behaviours (Bickerstaff and Walker, 2005).

Yet interdependencies between actors shift, and power relations alter. New discourses generate new expectations about the adequacy of regime performance (such as its sustainability) and contribute to a re-ordering of priorities (Fouilleux, 2000; Muller, 2000). The status of resources and the regime position of different actors are cast in a new light. Shifts in relations of power therefore need careful attention in transition studies (Duineveld *et al.*, 2009). Indeed, politics is the constant companion of transitions, serving as context, arena, obstacle, enabler, arbiter and manager of repercussions (Meadowcroft, 2011).

Thus, for a niche to break through and a (potential) transition to emerge, requires niche actors to develop a political capacity for positioning the niche favourably in the light of ongoing processes (e.g. environmental or economic crises), mobilizing support, influencing agendas and re-directing investments and policy commitments away from incremental repair work and towards a more radical transition. Indeed, regime transformations are an emerging outcome of resource-interdependent actors negotiating material responses to future expectations (Stirling and Smith, 2008).

When analysing case studies it is useful to identify the strategies employed by various actors to instigate a societal change process, and strategies used by regime actors to ensure stability. Such strategies may include lobbying; formation of networks, coalitions and alliances; playing the media; use of rhetoric; selective use of the results of scientific research; funding specific types of research; selecting specific stakeholders for inclusion in participatory processes; making and implementing laws, formal rules and procedures; or transforming institutions (Duineveld *et al.*, 2009).

Sustainability: politics and definitions

One of the key issues in transitions to sustainability is the process involved in identifying which problems need to be addressed and selecting suitable approaches to address them. This is a political, constructed and often contested process. Indeed, there is typically ample scope for debate over the sustainability of both incumbent regime and alternative niches (Stirling and Smith, 2008). Sustainability appraisals are necessarily undertaken from different positions and perspectives. Overall goals for sustainability, such as preservation of biodiversity or reducing the environmental impact of agricultural practices, often achieve broad rhetorical consensus. However, more specific criteria tend to be hotly contested with profound implications for the favoured pathways. A typical example is the current debate regarding the sustainability of biofuel production, which is rife with ambiguities on the choice of indicators, the projected future environmental and societal impact, and the relative weighing of effects in developed and emerging countries.

Regarding agriculture, there are obviously several contending paradigms (van der Ploeg, 2009; Freibauer *et al.*, 2011; Kitchen and Marsden, 2011; Levidow, 2011). There is ample discussion, for example, on whether a transition to sustainability can be achieved by focusing on technological artefacts (e.g. GMOs, nanotechnology, precision agriculture) or whether it is more effective to focus on consumer behaviour, social relations, allocation rights, institutional structures and cultural perspectives. Each of these elements is part of a discourse, and there is intense debate as to which standards are suitable, as well as which criteria adequately reflect sustainability. Thus, transitions both presuppose and bring about a shift in standards of legitimacy.

These standards of legitimacy are reflected in the conceptual frames that define which problems are persistent (while ignoring and downplaying others), and which solutions are appropriate to address the problems. As a result, emerging transitions tend to be rooted in contrasting sets of interests and prospects, different values, and cognitive frames. A societal discourse ensues, often by influential members of the established regime (such as agri-business groups, banks, state agencies, expert systems or researchers). Regime actors are likely to attempt to block transitions that are advocated as necessary by particular lobbies, and to support others by arguing that they are 'objectively necessary', given the rationality of their cognitive frame (van der Ploeg, 2009). However, what counts as 'authoritative knowledge' is often as much a reflection of institutional power as it is of robust or comprehensive understanding (Stirling, 2009). The issue of the definition of what counts as a transition to sustainability is therefore closely related to the question of 'whose system counts' (Stirling, 2008), which includes the definition of the boundaries of the system under consideration as well as what its structure is and how it functions (Shove and Walker, 2007). Thus, the identification of persistent problems, the choice of criteria to assess the relative worth of alternative pathways, and the solutions that are understood as leading to sustainability, are all the result of social interaction, political decision-making and conflict.

Institutions: rules, values and lifestyles

Another aspect that has not stood at the centre of previous studies on transitions to sustainability is the role of institutions. Institutions are all those 'rules of the game' such as norms, conventions and ways of doing things that structure human interaction and activity (North, 2005). Rules can be formal or informal, overt or implicit. Rules are expressed in artefacts, such as long-lived material infrastructures. Socially agreed rules of interpretation and signification of the external world also build cognitive frameworks. These cognitive frameworks are embodied in discourse and narratives through which people make sense of their environment. Similarly, lifestyles are the embodiment of societal conventions and values. Artefacts, cognitive frameworks and lifestyles tend to dampen, delay and raise the stakes of attempts at rule reformulations.

Essentially, transitions involve changing the incentives for how individuals and organizations behave, which in turn means changing institutions (Woodhill, 2009). Indeed, incentives for behaviour themselves come from a complex and highly interconnected web of institutional factors, not least of them belief systems (North, 2005).

Conclusion

This book covers a variety of case studies. Some of these case studies are still 'novelties', or niches in the process of establishment; others are niches that have matured and have gained momentum. This diversity allows the analysis of various learning processes; how niche and regime actors have navigated complexity and unpredictability and how they have built (more or less) stable networks. In some cases, the values of a niche have become more broadly accepted, and in the process their networks have become larger and now include powerful regime actors that secure resources for the niche, and may convey legitimacy to the solutions they propose to a sustainability problem.

The focus of the analysis is a better understanding of niche-regime interactions, rather than reconstructing long-term transitional pathways. Indeed, whether or not a niche will 'break through' and 'take off', whether it will effectively initiate a transition, is outside the scope of the analysis since it would require a longitudinal study; a historical retrospective. The analyses emphasize technological, network and institutional anchoring processes. The aim is to better understand learning processes, network dynamics and struggles between niche and regime actors. This understanding will shed light on how niche (and regime) actors take advantage of regime-internal tensions, how niche actors build linkages with specific regime actors to support their niche, or how they create new relationships between regimes.

The comparative analysis of case studies from several countries also allows the identification of institutional arrangements, governance approaches and supporting measures that have helped the niche to initiate change at regime level, taking into account the specificities of the context, including pressures from the socio-technical landscape, framing of persistent problems, involved actors, cultural traditions, and historical developments. The analysis thus highlights the context-dependence and unpredictability of change processes in farming.

References

Barbier, E. (2011) Transaction costs and the transition to environmentally sustainable development. *Environmental Innovation and Societal Transitions* 1, 58–69.

Belz, F.-M. (2004) A transition towards sustainability in the Swiss agri-food chain (1970-2000): Using and improving the multi-level perspective. In: Elzen, B., Geels, F. and Green, K. (eds) *System Innovation and the Transition to Sustainability*. Edward Elgar Publishing, Cheltenham, UK, pp. 97–113.

Bickerstaff, K. and Walker, G. (2005) Shared visions, unholy alliances: Power, governance and deliberative processes in local transport planning. *Urban Studies* 42, 2123–2144.

Bos, B. and Grin, J. (2008) 'Doing' reflexive modernization in pig husbandry: The hard work of changing the course of a river. *Science, Technology and Human Values* 33, 480–507.

de Haan, J. and Rotmans, J. (2011) Patterns in transitions: Understanding complex chains of change. *Technological Forecasting and Social Change* 78, 90–102.

Diaz, M., Darnhofer, I., Darrot, C. and Beuret, J-E. (2013) Green tides in Brittany: What can we learn about niche-regime interactions? *Environmental Innovation and Societal Transitions* 8, 62–75.

Duineveld, M., Beunen, R., van Assche, K., During, R. and van Ark, R. (2009) The relationship between description and prescription in transition research. In: Poppe, K., Termeer, C. and Slingerland, M. (eds) *Transitions Towards Sustainable Agriculture and Food Chains in Peri-Urban Areas*. Wageningen Academic Publishers, Wageningen, the Netherlands, pp. 309–323.

Elzen, B., Geels, F.W. and Green, K. (2004a) Conclusion. Transitions to sustainability: Lessons learned and remaining challenges. In: Elzen, B., Geels, F. and Green, K. (eds) *System Innovation and the Transition to Sustainability. Theory, Evidence and Policy*. Edward Elgar Publishing, Cheltenham, UK, pp. 282–300.

Elzen, B., Geels, F. and Green, K. (eds) (2004b) *System Innovation and the Transition to Sustainability. Theory, Evidence and Policy*. Edward Elgar Publishing, Cheltenham, UK.

Elzen, B., Geels, F., Leeuwis, C. and van Mierlo, B. (2011) Normative contestation in transitions in the making: Animal welfare concerns and system innovation in pig husbandry (1970-2008). *Research Policy* 40, 263–275.

Elzen, B., van Mierlo, B. and Leeuwis, C. (2012) Anchoring of innovations: Assessing Dutch efforts to harvest energy from glasshouses. *Environmental Innovation and Societal Transitions* 5, 1–18.

Fouilleux, E. (2000) Entre production et instutionalisation des idées. La réforme de la politique agricole commune [Between production and institutionalisation of ideas. The reform of the CAP]. *Revue Française de Science Politique* 50, 277–306.

Freibauer, A., Mathijs, E., Brunori, G., Damianova, Z., Faroult, E., Girona i Gomis, J., O'Brien, L. and Treyer, S. (2011) *Sustainable Food Consumption and Production in a Resource-constrained World*. 3rd SCAR Foresight Exercise. DG for Research and Innovation, Biotechnologies, Agriculture, Food.

Geels, F.W. (2002) Technological transitions as evolutionary reconfiguration processes: A multi-level perspective and a case-study. *Research Policy* 31, 1257–1274.

Geels, F.W. (2004) From sectoral systems of innovation to socio-technical systems: Insights about dynamics and change from sociology and institutional theory. *Research Policy* 33, 897–920.

Geels, F.W. (2005) The dynamics of transitions in socio-technical systems: A multi-level analysis of the transition pathway from horse-drawn carriages to automobiles. *Technology Analysis and Strategic Management* 17, 445–476.

Geels, F.W. (2009) Foundational ontologies and multi-paradigm analysis, applied to the socio-technical transition from mixed farming to intensive pig husbandry (1930-1980). *Technology Analysis and Strategic Management* 21, 805–832.

Geels, F.W. (2011) The multi-level perspective on sustainability transitions: Responses to seven criticisms. *Environmental Innovation and Societal Transitions* 1, 24–40.

Geels, F.W. and Schot, J. (2010) The dynamics of transition: A socio-technical perspective. In: Grin, J., Rotmans, J. and Schot, J. *Transitions to Sustainable Development. New Directions in the Study of Long Term Transformative Change*. Routledge, New York, pp. 9–101.

Giddens, A. (1984) *The Constitution of Society: Outline of the Theory of Structuration*. Polity Press, Oxford, UK.

Grin, J. (2010) Modernization processes in Dutch agriculture 1886 to the present. In: Grin, J., Rotmans, J. and Schot, J. *Transitions to Sustainable Development. New Directions in the Study of Long Term Transformative Change*. Routledge, New York, pp. 249–264.

Grin, J., Rotmans, J. and Schot, J. (2010) *Transitions to Sustainable Development. New Directions in the Study of Long Term Transformative Change*. Routledge, New York.

Hermans, F., van Apeldoorn, D., Stuiver, M. and Kok, K. (2013) Niches and networks: Explaining network evolution through niche formation processes. *Research Policy* 42, 613–623.

Holtz, G., Brugnach, M. and Pahl-Wostl, C. (2008) Specifying 'regime' – a framework for defining and describing regimes in transition research. *Technological Forecasting & Social Change* 75, 623–643.

Kemp, R. and van Lente, H. (2011) The dual challenge of sustainability transitions. *Environmental Innovation and Societal Transitions* 1, 121–124.

Kitchen, L. and Marsden, T. (2011) Constructing sustainable communities: A theoretical exploration of the bio-economy and eco-economy paradigms. *Local Environment* 16, 753–769.

Lamine, C. and Bellon, S. (eds) (2009) *Transitions vers L'agriculture Biologique* [Transitions towards Organic Farming]. Editions Quæ, Versailles.

Levidow, L. (ed.) (2011) *Agricultural Innovation: Sustaining What Agriculture? For What European Bio-economy?* Final report of the project 'Co-operative Research on Environmental Problems in Europe' (CREPE). Available from: http://crepeweb.net.

Marsden, T. (2013) From post-productivism to reflexive governance: Contested transitions in securing more sustainable food futures. *Journal of Rural Studies* 29, 123–134.

Meadowcroft, J. (2011) Engaging with the politics of sustainability transitions. *Environmental Innovation and Societal Transitions* 1, 70–75.

Milone, P. (2009) *Agriculture in Transition. A Neo-institutional Analysis*. Van Gorcum, Assen, the Netherlands.

Muller, P. (2000) L'analyse cognitive des politiques publiques: Vers une sociologie politique de l'action publique [A cognitive analysis of public policies: Towards a political sociology of public action]. *Revue Française de Science Politique* 50, 189–208.

North, D. (2005) *Understanding the Process of Economic Change*. Princeton University Press, Princeton, New Jersey.

Oostindie, H. and van Broekhuizen, R. (2008) The dynamics of novelty production. In: van der Ploeg, J-D. and Marsden, T. (eds) *Unfolding Webs. The Dynamics of Regional Rural Development*. Van Gorcum, Assen, the Netherlands, pp. 68–86.

Poppe, K., Termeer, C. and Slingerland, M. (eds) (2009) *Transitions towards Sustainable Agriculture and Food Chains in Peri-Urban Areas*. Wageningen Academic Publishers, Wageningen, the Netherlands.

Rip, A. (2006) A co-evolutionary approach to reflexive governance – and its ironies. In: Voss, J-P., Bauknecht, D. and Kemp, R. (eds) *Reflexive Governance for Sustainable Development*. Edward Elgar Publishing, Cheltenham, UK, pp. 82–100.

Roep, D. and Wiskerke, J. (2004) Reflecting on novelty production and niche management in agriculture. In: Wiskerke, J. and van der Ploeg, J-D. (eds) *Seeds of Transition. Essays on Novelty Production, Niches and Regimes in Agriculture*. Van Gorcum, Assen, the Netherlands, pp. 341–356.

Schot, J. and Geels, F. (2008) Strategic niche management and sustainable innovation journeys: Theory, findings, research agenda, and policy. *Technology Analysis and Strategic Management* 20, 537–554.

Schmid, O., Sanders, J., Schermer, M. and Hamm, U. (2004) Organic marketing initiatives and rural development: Conclusions and policy recommendations. In: Schmid, O., Sanders, J. and Midmore, P. (eds) *Organic Marketing Initiatives and Rural Development*. OMIaRD, Vol. 7. School of Management and Business, University of Wales Aberystwyth, pp. 181–200.

Shove, E. and Walker, G. (2007) Caution! Transitions ahead: Politics, practice and sustainable transition management. *Environment and Planning A* 39, 763–770.

Sinclair, K. (2014) Transformative change in contemporary Australian agriculture. PhD thesis. School of Environmental Sciences, Charles Sturt University.

Slingerland, M. and Rabbinge, R. (2009) Introduction. In: Poppe, K., Termeer, C. and Slingerland, M. (eds) *Transitions Towards Sustainable Agriculture and Food Chains in Peri-Urban Areas*. Wageningen Academic Publishers, Wageningen, the Netherlands, pp. 13–23.

Smith, A. (2006) Green niches in sustainable development: The case of organic food in the United Kingdom. *Environment and Planning C: Government and Policy* 24, 439–458.

Smith, A. (2007) Translating sustainabilities between green niches and socio-technical regimes. *Technology Analysis and Strategic Management* 19, 427–450.

Spaargaren, G., Loeber, A. and Oosterveer, P. (eds) (2012) *Food Practices in Transition. How Globalization and Sustainable Development are Changing the Consumption, Retail and Production of Food*. Routledge, Oxford, UK.

Stirling, A. (2008) Opening up and closing down: Power, participation and pluralism in the social appraisal of technology. *Science, Technology and Human Values* 33, 262–294.

Stirling, A. (2009) *Direction, Distribution and Diversity! Pluralising Progress in Innovation, Sustainability and Development*. STEPS Working Paper 32. STEPS Centre, Brighton, UK.

Stirling, A. and Smith, A. (2008) *Social-ecological Resilience and Socio-technical Transitions: Critical Issues for Sustainability Governance*. STEPS Working Paper 8. STEPS Centre, Brighton, UK.

Tisenkopfs, T., Brunori, G., Knickel, K. and Sumane, S. (2009) Co-production of rural innovation: Towards an enriched theoretical model. In: Knickel, K., Tisenkopfs, T. and Peter, S. (eds) *Innovation Processes in Agriculture and Rural Development. Results of a Cross-national Analysis of the Situation in Seven Countries, Research Gaps and Recommendations*. Report of the IN-SIGHT project. Available from: www.insightproject.net

van der Ploeg, J.D. (2009) Transition: Contradictory but interacting processes of change in Dutch agriculture. In: Poppe, K., Termeer, C. and Slingerland, M. (eds) *Transitions towards Sustainable Agriculture and Food Chains in Peri-Urban Areas*. Wageningen Academic Publishers, Wageningen, the Netherlands, pp. 293–307.

van der Ploeg, J.D. and Marsden, T. (eds) (2008) *Unfolding Webs. The Dynamics of Regional Rural Development*. Van Gorcum, Assen, Netherlands.

Vanloqueren, G. and Baret, P. (2009) How agricultural research systems shape a technological regime that develops genetic engineering but locks out agroecological innovations. *Research Policy* 38, 971–983.

Vellema, S. (ed.) (2011) *Transformation and Sustainability in Agriculture. Connecting Practice with Social Theory*. Wageningen Academic Publishers, Wageningen, Netherlands.

Vogelezang, J., Wals, A., van Mierlo, B. and Wijanads, F. (2009) Learning in networks in Dutch agriculture: Stimulating sustainable development through innovation and change. In: Poppe, K., Termeer, C. and Slingerland, M. (eds) *Transitions towards Sustainable Agriculture and Food Chains in Peri-Urban Areas*. Wageningen Academic Publishers, Wageningen, the Netherlands, pp. 93–111.

Wilson, G. (2007) *Multifunctional Agriculture: A Transition Theory Perspective*. CABI, Wallingford, UK.

Wiskerke, J. and van der Ploeg. J.D. (2004) *Seeds of Transition: Essays on Novelty Production, Niches and Regimes in Agriculture*. van Gorcum, Assen, the Netherlands.

Woodhill, J. (2009) Institutional innovation and stakeholder engagement. In: Poppe, K., Termeer, C. and Slingerland, M. (eds) *Transitions towards Sustainable Agriculture and Food chains in Peri-Urban Areas*. Wageningen Academic Publishers, Wageningen, the Netherlands, pp. 273–291.

Chapter 3

Understanding the diversity of European rural areas

B. Slee[1], T. Pinto-Correia[2]

[1]*James Hutton Institute, Aberdeen (bill.slee@hutton.ac.uk);* [2]*ICAAM, University of Évora*

Introduction

In this chapter, we first sketch out the structure and diversity of European agriculture and its evolution to its present state. Second, we examine the assembled evidence that explores differences across Europe in the way the farm sector has evolved, offering tentative reasons as to causes of difference and patterns of change. Third, we set these farm-level changes within a wider framework of understanding the relationship between agricultural change and broader economic change, addressing societal demands on the rural space. Finally, we reflect briefly on the challenges of managing transitions in the farm sector given its diversity and non-homogeneity. The relationships we explore relate partly to the interactions between agriculture and the economy and also partly to the policies that shape those interactions. Such information may help frame our overall understanding of evolutionary pathways and sustainability transitions in the rural economy.

An exploration of existing and possible future pathways towards sustainability in European farming can be framed in recognition of the diversity of rural Europe. Three types of diversity are of particular importance in this respect. First, there is diversity in production systems, which is largely, though not exclusively, conditioned by the biophysical potential of land. Second, there is diversity in the structural characteristics of farming and other land use activity and enterprise. Third, there is great diversity in socio-economic conditions, not just within the rural land use sector, but also within the regional economies of which the rural sector is but part. In the latter two forms of diversity, path dependencies play a huge part in shaping aggregate socio-economic character, structural characteristics of farming and the future opportunities for transitions towards more sustainable agricultures.

Ignoring anomalous overseas non-European territories which are part of the EU, European rural land use embraces Mediterranean conditions from Portugal in the west, to Cyprus in the east and, in a more northerly transect, temperate grassland in Ireland, to the boreal forests of North Karelia in Finland. Large mountain ranges impact significantly on rural land use, with the Pyrenees, Massif Central, the Alps and the Carpathians constituting a watershed that virtually spans Europe from west to east with other upland and mountain regions from Greece to Portugal to Scotland and Sweden also creating limitations on, and thereby structuring, land use possibilities. Operating within these biophysical parameters, there is a diverse range of farming and rural land use systems. Northern European agriculture hinges around grassland and ruminant livestock systems with grain production where conditions permit (more to the east than the west). In southern Europe, tree (and vine) crops on farms assume greater importance in shaping what remains a more regionally grounded set of systems of farming and food production, although grassland, grain crops and ruminants are still of great importance. Pigs and poultry are found throughout Europe, with highly variable systems of production from extensive systems in the *dehesa* and *montado* of Iberia (under silvo-pastoral *Quercus ilex* and *Quercus suber* systems) to intensive indoor systems in Denmark.

In some regions, agriculture has a lengthy history of long-distance trade; in others, agricultural output has delivered primarily to local and regional economic demands. Even in the pre-industrial period there was trade in agricultural products over considerable distances. Wine was exported across Europe; as was wool, particularly high-quality wool which produced great prosperity in both production and manufacturing areas. Even where agricultures were predominantly subsistence-based innovation still occurred; such as the slow but eventually widespread adoption of the potato across northern and western Europe (Salaman, 1965); and the selection of regionally adapted varieties of crop and animal. The nature and extent of major changes in farming – the so called 'agricultural revolutions' – have been much debated. Technical innovations of various types, including mechanization, land drainage techniques, feeding and breeding have occurred with varying levels of uptake over space but have tended to occur in pulses which some have labelled revolutions. Some might argue that contemporary science built around genetic manipulation and cell biology places us on the threshold of yet another revolution. Notwithstanding such pulses of innovation, by the mid-20th century livestock breeds were still highly concentrated regionally and European agriculture retained an intensely regional character, evidenced in the wide variety of regional foods found in the more southern, peripheral and disadvantaged parts of Europe. It may be pertinent to think in terms of a core area, in which new technology is readily adopted, and a periphery where more complex processes of deconstruction and reconstruction of knowledge take place to shape farming practices and products (van der Ploeg and Long, 1994). Indeed, innovations involved in deconstruction and reconstruction of core knowledge are often the means of developing sustainable agricultures in more biophysically compromised areas (Ortiz-Miranda *et al.*, 2013). In such areas, what may have been lacking are the multiple processes of social and institutional innovation required for technical innovations to break through (Moragues-Faus *et al.*, 2013).

Over the last thousand years, a continent which had been characterized by a largely feudal agricultural structure has undergone considerable but uneven change with multiple consequences. In some regions, the small peasant farm structure remains, often replaced on better land by more capitalist forms of farming. However, in the relatively fertile areas of Europe, especially in the regions surrounding the triangle from Frankfurt to London to Paris, capitalist farming replaced peasant farming at a relatively early stage – certainly by the early 19th century as the transformative effects of the industrial revolution began to increase demand for food and this, in turn, stimulated changes in the farm sector. In contrast, large parts of central and eastern Europe experienced 50 years or more of socialist collectivization from the early- to mid-20th century: another form of industrial agriculture. After the collapse of that experiment in the early 1990s, collective agriculture was often largely 're-peasantized' by restitution policies and practices more intent on social justice than creating commercially viable farms. Islands of commercial, large-scale capitalist farming have since found a niche in some post-socialist areas where restitution was to farms that had capacity to become commercial; where consolidation of holdings has occurred; or where state farms were bought up by capitalist farming ventures.

For a variety of reasons, European agriculture was perhaps not as wholly penetrated by capitalist logic as some commentators, such as Newby *et al.* (1978) might have presumed. Others have argued against this view. Family farming has long been a significant part of European agriculture (Gasson *et al.*, 1988; Gasson and Errington, 1993). Harriet Friedmann's (1980) arguments about simple commodity production and its differences from capitalist commodity production may have been heavily criticized by some (e.g. Goodman and Redclift, 1985) but there is something compelling about the durability of what can only be described as family businesses in the farm sector. Much earlier, Chayanov (1966) had made the same point about the peasant farm household's capacity for survival. Agriculture does not change evenly across regions, farm types or across tenancy and land ownership structures. In many areas there remains a strong legacy of small farms, many not capable of supporting a household (Ortiz-Miranda *et al.*, 2013).

In a European context, the work of Marsden *et al.* (1996) probes further into farming's increasing engagement in the market through processes of subsumption. More recently, Potter and Tilzey (2005) have argued that this is more than a market process and is substantially assisted by neoliberal policy architecture. Family farms may not be wholly capitalist entities, but they are increasingly locked into circuits of capital in agri-food systems which shape both development trajectories and sustainability outcomes. Potter and Tilzey (2005) further argue that support for this transition towards incorporation into major circuits of capital in the agri-food system drives policy, rather than post-productivist values or multifunctionality. We return to such issues below.

We should not accept the European idealization of the family farm uncritically. Alanen (2002) argues that the desire for reinstatement of the family farm in post-socialist countries through restitution is premised on ideological grounds unrelated to the desire to produce a competitive, market-oriented farm sector. Furthermore, from the early 1990s onwards, Bryden's (1988) work has also shown that, in many ways, family farms are embedded in a wider economy through pluriactivity, not just the agri-food complex, even where they are still significantly self-sufficient. The much-heralded independence of the family farm is thus illusory because it is often sustained by income derived from other sectors of the economy and a locking-in to wider circuits of agri-food capital. Indeed, the connectivity between members of farm households and the economy may be a vital force in maintaining small farm family units.

A further complication in the rural land use sector is the extent to which farming is not always intended to comprise formal market driven economic activity, not only because of subsistence demands but also because of recreational preferences of more affluent farm owners (Primdahl and Kristensen, 2011). Hobby farms (or lifestyle farms) have been identified as a significant proportion of farming activity in many regions of Europe (Pinto-Correia *et al.*a, this volume). Hobby farmers typically have a significant external (non-farm) income source and tend to treat their farms as platforms for recreational activity, which may or may not have marketable outputs. The range of hobby farming types is enormous. In northern Europe, such farms are often used as places for recreational horses and, particularly in the horse sector, the boundary between commercial and lifestyle activity can be very difficult to distinguish. In other regions of Europe, hobby farming is often dedicated to vegetables and fruits but olive oil and vines often provide a recreational activity that can also be semi-commercial at times (Arnalte-Alegre and Ortiz-Miranda, 2013; Pinto-Correia *et al.*, 2014).

The third source of diversity which frames European farm adjustment is that created by proximity to other forms of economic activity. Over 200 years ago, von Thünen became acutely aware of the impact of an urban centre on the intensity and type of surrounding land use. The geography of the industrial revolution based first around water power and then coal and steel production in 19th century Europe, created powerful demand hubs in the Ruhr, central and northern England, south Wales, central Scotland and later in north-east Spain and northern Italy. The emergence of these hubs had huge impacts on regional labour markets, drawing in a workforce from the surrounding countryside; but they also created a demand for food, met partly by local supply and, increasingly in the case of the United Kingdom, by imports from countries such as Canada, Australia and New Zealand. Over time and with enhanced mobility, their rural hinterlands provided rural retreats for wealthy industrialists and others, particularly in high amenity landscapes and who often entered farming as much for leisure as for profit.

By the new millennium, de-industrialization and new technologies had begun to reshape the nature of urban centres. The service economy has replaced manufacturing as the dominant force in post-industrial urban development with its own geography of concentration and it is this, built on a legacy of medieval market towns and cities and industrial revolution decay and urban regeneration, that creates the patterns of contemporary population in Europe. Although some rural areas retain many features of socio-economic decline, others have become highly prosperous with some of the highest living standards in Europe.

As well as taking the farm household as the unit of analysis it is possible to hypothesize different types of countryside and explore change at a regional or sub-regional scale. Marsden *et al.* (1993) proposed a four-fold classification of types of rural area which included the following categories: productivist countryside; contested countryside; preserved countryside; and paternalistic countryside. Regional (or sub-regional) agriculture will arguably be shaped by the dominance of one type of countryside. These are ideal types rather than observable realities on the ground but they encapsulate the diversity of pressure and process within an increasingly differentiated countryside. Holmes (2006, 2012) and Copus and Hörnström (2011) offer a more contemporary take on types of region and have mapped such areas using socio-economic datasets.

The policy context

Contemporary European agriculture is not just a product of food and fibre markets and leisure consumption. Its development trajectory has been significantly mediated by policies. Initially, those policies evolved primarily around agriculture but more recently developments in the farm sector are increasingly mediated by changes in the wider economy. Further, environmental policy and spatial and regional planning policy have become growing influences on agriculture. Some would also argue (Peters, 1991; Primdahl and Swaffield, 2010) that macroeconomic policies also impact greatly on the farm sector.

This is evidenced in the retreat to the land during the austerity years after the collapse of the planned economies of eastern and central Europe in the early 1990s and is evidenced again in the flight to the land in countries such as Greece and Spain in the wake of the recent (2008 and subsequently) European financial crisis. Currency fluctuations have also impacted significantly on the profitability of farming in member states without the Euro.

Most of continental Europe turned strongly protectionist towards the farm sector in the aftermath of major 19th and 20th century wars. Self-sufficiency was a compelling argument under the prospect of a war of attrition or a blockade. However, the UK, as the first industrial nation, took a different view, abandoning protectionism for a century from 1846 and opting for free trade to feed its growing urban population cheaply and maintain industrial competiveness (Tracy, 1989). The expansion of the European Union (EU) led to an end to UK policy exceptionalism and continuity in, and deepening of, the continental protectionist model of support. This was partially reduced by agriculture's entry into the General Agreement on Tariffs and Trade (GATT) and the World Trade Organization (WTO) in the 1990s but by then, agricultural support was soaking up over half of the total European Community's budget. In response to the WTO agreement, farm policy has changed but the farm sector still receives the lion's share of public support to the rural sector, rather than the wider rural economy. This level of public expenditure may not be sustainable financially, let alone with respect to its environmental consequences (see below).

The early years of the new millennium have been marked by considerable price volatility in global food markets. The reduced protection arising from WTO compliance has exposed European farmers to this global price volatility to an unprecedented degree compared to the earlier protectionist embrace of the Common Agricultural Policy (CAP). At the same time, these increases in food prices have led to a reappraisal of the importance of production of food and fibre in the rural economy and sparked debate about the possibility of sustainable intensification. Rising food prices have also thrown a spanner in the works of the emerging European bio-economy. Both globally and in the EU, policy targets for biofuel production look set to exacerbate food price rises as well as generating a further set of environmental problems, and both member states and the EU have pulled back from early targets for expansion, especially in relation to first generation biofuels.

Another facet of European policy support has been the sustenance of product variety and traditional practices using European legislation. This is not old style frontier protectionism but the protection and support of traditional practices, thereby adding to the diversity and multifunctionality of European agriculture. Such protection as afforded for Protected Destination of Origin (PDO) or Protected Geographical Indication (PGI) status can be seen as state-supported means of sustaining diversity and protecting the interests of particular groups of producers. Many would argue that this protected status creates a platform on which more sustainable agricultural futures can be built (Sjoblom *et al.*, 2012).

Over time, the farm sector has had to cope with burgeoning environmental policies, both at nation-state and EU levels. Agriculture's implication in biodiversity decline, landscape quality deterioration and water quality concerns has been increasingly met by new regulation. The European Habitats and Water Framework Directives limit the room for manoeuvre in farm business adjustment. This environmental regulation and the panoply of support emanating from the second pillar of the reformed CAP now interact with markets and changing social attitudes and preferences to frame the adjustment pathways chosen.

Another environmental arena where regulation will increasingly impact on the farm sector is in relation to societal responses to climate change. The EU has instituted policy targets with respect to decarbonization in the Europe 2020 initiative and expects these to be embedded in Rural Development Programme actions. Under the 'Resource Efficient Europe' Flagship Initiative, the European Commission (EC) seeks to "support the shift towards a resource efficient and low-carbon economy that is efficient in the way it uses all resources. The aim is to decouple our economic growth from resource and energy use, reduce CO_2 emissions, enhance competitiveness and promote greater energy security" (EC, 2010:15). It is widely recognized that the farm sector is both instrumental in significant emissions and has capacity to sequester carbon in various ways.

A further area of engagement with wider public policies is that relating the farm sector to spatial planning. There has been a retreat from thinking sectorally about agriculture towards thinking more in terms of territorial development. There is recognition that agricultural developments are often framed by regional and sub-regional influences, particularly proximity to large cities. Forward-looking cities (such as Amsterdam, or Rennes, the example used in Darrot *et al.*, this volume) are increasingly looking to develop more integrated approaches to planning low carbon futures with their hinterlands exploited for sustainable local food and rural recreational opportunity. Further, farm diversification has often thrown farm businesses into the more uncertain regulatory structures of town and country planning, when regulatory approval may be required for tourism or food processing developments (see Curry and Owen, 1996 for an early discussion of this). Overall, when engaging in diversification or maintenance of less productive farm systems which secure highly valued ecosystem services, land managers are often confronted with conflicting policy discourses and tools, leading to difficult tensions in the everyday decisions (Pinto-Correia *et al.*, 2014). Much more integration across sectors at the level of policy design and at the level of the local administration can be seen as needed to support sustainability transitions (see Pinto-Correia *et al.*b and Darnhofer *et al.*, both this volume).

How do we interpret these policy changes? In the 1990s, a number of authors adopted a post-productivist interpretation of policy change and, in spite of strong interest from some (Mather *et al.*, 2006), the concept has often been criticized as unclear (Wilson, 2001; Evans *et al.*, 2002). Indeed, Potter and Tilzey (2005) strongly reject post-productivism and argue that the advance of market logic is aided by a neoliberal policy discourse which prevails over competing policy discourses. The less controversial term 'multifunctional' has increasingly been used by both policy makers and academics to describe the principles and policies supporting agriculture in the EU. However, the neoliberal policy agenda implies a primary policy ethos to support the agri-food system, rather than deliver to the multifunctionality agenda and support public good delivery within a post-productivist/ multifunctional policy framework. Holmes' (2006, 2012) re-diagnosis of post-productivism as multifunctionality partly overcomes the Potter and Tilzey critique but in relation to the European context, perhaps fails to disentangle the struggle between support for farming enterprise (which, as Pillar 1, clearly remains the dominant part of the CAP) and support for public goods (and thereby multifunctionality) which is effected largely through the much weaker Pillar 2 of the CAP (i.e. the Rural Development Programme (RDP)). Whilst the multifunctionality policy discourse is powerful, it arguably remains subsidiary to the overarching support of production and, in the shadow of food security debates, it may even be possible to argue for a post-post-productivist productivism, or more simply neo-productivism or a new bio-economy (Marsden, 2013).

Diversity of production

In the early years of 'the European project' the farm sector was a central part of the contract and it remains so today, certainly in terms of EU spending and arguably in the wider politics of the EU. In the 1960s, much of European agriculture was regarded as backward and inefficient, dominated by small farms. An early Agricultural Commissioner, Sicco Mansholt, sought to modernize farming and emancipate it from the small and inefficient structures that remained in the so-called 'Mansholt Plan'. Later, he recanted on his earlier vision as he became more aware of the consequences of the modernization project on the environment.

In spite of various policies to support modernization, there remain large variations in structures and production systems across Europe and over time, some differences have tended to be reinforced by increased specialization. Some mixed farming areas remain but the core arable areas have tended to lose livestock (except pigs and poultry) and specialist livestock production has become more intensive. Once agriculture was emancipated from the need to sustain fertility 'internally' through the on-the-farm use of rotations and manures, first through import of animal fertilizers (guano) and later artificial nitrogen and mined phosphate and potash, specialization could be more readily accomplished and economies of size and scale delivered. This specialization might have been hindered by ecological consequences of concentration, such as pests and diseases, but a developing arsenal of agri-chemicals was now available. Sometimes, relief was temporary as weeds and pathogens built up resistance. But modernization enabled significant yield increases in core areas, at the same time as locking farmers into high-input pathways.

However, intensification was not without consequence. Landscape structures were simplified to allow for larger machines; agri-chemicals and their residues began to have a discernible effect on biodiversity and sometimes water quality; phosphates and nitrates began to have an adverse effect on water quality; and concentrations of animal production caused major challenges in the disposal of manure.

Tolerance of diversity, however, has increased since the modernization project was promoted during the Mansholt years. Even semi-subsistence farming can now receive special support under the RDP. But there are inevitable tensions between agricultures geared for globalized commodity chains, and agricultures geared to semi-subsistence and local markets (Primdahl and Swaffield, 2010). This duality seems likely to produce a variegated Europe of specialized food production regions and a periphery in which diverse regionally specific production practices remain, sometimes as in the Tuscan model, benefiting from tourism and new forms of rural occupancy, sometimes still reliant on regional and local markets.

Diversity of farm structures and incomes

Farm size varies significantly across Europe, though business size cannot always be equated with areal extent as there are many extensively farmed areas where farms are territorially large but remain small business. Tenure also varies, although the collapse of state and collective ownership in Eastern Europe has confirmed the dominance of private property tenure. In some areas, tenanted farmland is still a common feature, and the dynamics of business adjustment result in individual businesses expanding through very different forms of tenure (Lobley *et al.*, 2002). In many countries, there is a legacy of common property, from the *baldios* of Portugal to the extensive commons of northern Scandinavia. Commons are found in the drylands and forests of the Mediterranean, Alpine pastures, and extensive grasslands in north-west Europe. New arrangements for shared land ownership or land management have also been emerging in recent years. For example, in Scotland, collective ownership opportunities have been created through new legislation and some communities now manage land collectively in a number of areas of north-west Scotland. However, any exploration of transition in European farming must recognize the prevailing context of family businesses and their decision-making practices.

Because farmers seldom retire and farming is dominated by family businesses, the farm sector has an ageing workforce. This is often deemed to be a problem because of the lack of new and young blood coming into the sector. Part of the explanation for this lies in the nature of the family business and the reluctance of farmers to retire, and in high land costs which prevent young people from breaking into the industry. Designing policy interventions to support new entrants, when the main cause of the lack of succession is the nature of family farms and high land prices, is politically popular but deeply problematic. Paradoxically, it is periods of recession in the farm sector, when land prices are cheaper and other land-based businesses fail thereby increasing the supply of land, which have probably been instrumental in attracting new blood into the industry, as it is then that the enthusiastic and motivated can gain access to farmland most cheaply.

Although there is a Common Agricultural Policy, farm incomes do not follow consistent trends across the EU, though a broad pattern is discernible. Eurostat data show that the top six performing countries in incomes per annual work unit in 2012 were all in new member states, from Romania in the south to Estonia in the north. Italy, Greece, Spain and Portugal were amongst the worst performing countries with real incomes from farming significantly below the European average. Ireland and Luxembourg were northern outliers with very low figures and the UK performed relatively well for north-west Europe. Given the significance of the Single Farm Payment, it seems likely that differences outside of the Euro-zone are a function of currency fluctuations. However, there would also appear to be other factors at work in shaping real incomes.

It is evident that there are major divergences in performance over a longer period of time and that the snapshot of figures in 2012 is reinforced by trends which show Mediterranean areas and Ireland performing rather badly, and most of the central and eastern European countries doing rather well, with only Slovenia amongst the post-socialist states not showing at least a 50% gain in farming income between 2007 and 2012. The principal lesson we can draw is that there do appear to be structural factors at work, although it is not clear what they are; and there are very significant differences between countries. These may have as much to do with national economies and macro-economic factors as with factors within the farm sector. Policy change is also important. Incomes seem not to be driven by structural factors such as farm size or the quality of fixed capital, except inversely, as relatively poor countries have done better in recent years. Of course, indices can suggest good performance when countries start from a low baseline and this may be the case for central and eastern European countries.

However, it has long been argued that rather than looking at farm incomes we should look at the total income of farm households. It is this bundle of income that determines household wellbeing, and the greater the extent of off-farm earning, the greater the total income of farm households. Such an observation also suggests that national figures on farm incomes are likely to mask total incomes of farm household figures between different types of farming and different regional economies, as there are known to be underlying fluctuations in sectors, but these tend to impact on only one part of farm household wellbeing.

Although farming has a major influence on shaping the rural landscape, it is no longer as important, economically, as it once was. Only in Portugal and many of the new eastern European member states does farm workforce exceed 10% of the total workforce; and it accounts for a markedly smaller proportion of GDP than its share of the workforce. It is not wholly clear whether the European financial crisis will increase farm employment in a temporary or more permanent way, as households engage with rural land in new ways. Both workforce and GDP shares have shown a tendency to decline significantly over time. Non-farm economic activities are thus likely to be major shapers of rural economic wellbeing; and regional wellbeing is likely to spill over into the farm sector through demand for hobby farms and rural recreation, and impacts on land and rural property prices.

There are also major intra-national variations in the extent of dependency on agriculture. Any region without a major urban hub is likely to have a much greater degree of primary sector dependency. Unless such areas have experienced a renaissance through tourism or rural re-population, they are more likely to be characterized by relatively fast population decline and very frail economies.

Exploring adjustment and pathways of change

Much of the early literature associated with the resurgent interest in farm adjustment from the 1970s onwards focused on the farm as the object of inquiry. This emphasis still exists in the work of authors such as Lobley *et al.* (2002) and remains a fertile research field. However, if the scale of observation moves from the individual farm to the region, it becomes necessary to see farm adjustment within a regional setting. As noted earlier, Marsden *et al.* (1993) had recognized that different types of countryside were evident but their four typologies can perhaps be seen more as Weberian ideals than discernible entities on the ground. More recently, various authors have tried to identify and map different areas using a range of multivariate techniques.

Studies on changes taking place in European rural areas suggest that there is a spatial, temporal and structural co-existence of several processes of transition, from increasingly specialized farm business to multifunctionality at both farm and landscape level (Woods, 2011; Pinto-Correia and Kristensen, 2013) and more recently to what Marsden (2013) calls bio-economic productivism, resulting in increasing diversification of rural space (Pinto-Correia and Breman, 2009; van Berkel and Verburg, 2011). While agriculture and its role as producer is still at the centre of a maelstrom of issues surrounding food safety, environmental balance and climate change, there is an increasing expectation from society that other goods and services will be provided by rural space (Selman, 2009; de Groot, 2010). The transformations at stake are connected partly to restructuring processes in the agricultural sector, as described in this chapter. They are also related to general trends of urbanization and regional economic performance. This includes urban sprawl and infrastructure development as well as changes in broad socio-economic processes. These changes result in the relocation of people and activities, in the sense both of a concentration in urban areas and a progressive emptying of remoter rural districts; and of counter-urbanization and new arrivals from the urban to the rural (Primdahl and Swaffield, 2010).

These emerging dimensions are linked to the involvement of a wider community of actors at multiple governance scales, increasing the social complexity of the rural (Marsden and Sonnino, 2008; Barbieri and Valdivia, 2010). In this way, multifunctionality and sustainability are not just an issue of diversification and adaptation in farming but can be interpreted as a paradigm shift in the management of rural space (Selman, 2009; Domon, 2011).

The multiple factors of change are combined in various ways (Wilson, 2007; Robinson, 2008) and occur in quite diverse directions and intensities in different regions and localities (van Berkel and Verburg, 2011). In the same location, divergent processes may have been taking place side by side, leading to complex patterns of occupancy, use of space, and economic activity (Short, 2008). This means that understanding processes taking place in the agricultural sector – and assessing pathways that may lead to enhanced sustainability in farming – cannot be evaluated in isolation of wider socio-economic change. Grasping the complex rural context, where agriculture is evolving within a wider set of regional restructuring processes, is now a fundamental component in the analysis of agricultural change.

This understanding of the broader rural context also raises the question of the farming sector not only as a production activity, but also including other land management activities. Land management may be not driven primarily by production or economic value but by environmental concerns, recreation goals, investments in amenity value and lifestyle – as the case studies in this book will show. These land management activities need to be acknowledged when studying farming adjustment pathways in Europe.

Whether post-productivism can be sustained under the threat of the so-called 'perfect storm' (Beddington, 2009) of climate change, increased energy prices, reduced food availability and water shortages is a moot point. Indeed, what Marsden (2013) terms bio-economic productivism has been widely embraced by the land-based community in a resurgent display of productivist rhetoric. Notwithstanding such rhetoric, rural Europe, which was already diversified, has become much less homogeneous, with diverging land management practices, products, and actors in different places, where practices are significantly shaped by regionally specific socio-economic factors. Former differentiation was grounded on the extremely diverse bio-physical conditions, structural factors, and different political and economic rationales. But the rural was more or less everywhere tuned by production agriculture as the dominant land use and the central economic and social role of agri-business. The present differentiation is much more complex, as it adds to the former but it is mainly related to divergent production paradigms, divergent land managers' motivations and multiple combinations of small-scale and global networks.

A substantial effort has been made already in creating and developing European-scale typologies for distinguishing between rural areas across Europe (van der Ploeg and Marsden, 2008; Copus and Hörnström, 2011; van Berkel and Verburg, 2011; van Eupen *et al.*, 2012). These typologies are mostly data-driven approaches undertaken by summing up data layers. Often, they do not address the new modes of rural dynamics and occupancies, which would require a stronger conceptual background (Holmes, 2006; Horlings and Marsden, 2011).

Copus and Hörnström (2011) have tried to structure different ruralities in the EDORA project, which addressed the need to challenge outdated generalizations about the nature of rural areas. Their framework recognizes both locational issues and structural characteristics in creating that diversity, resulting in a differentiation between agrarian, consumption and mixed countryside.

Other authors have described how rural space, from being essentially a production and living space some decades ago, has progressively also become recognized as a space of consumption and conservation (Woods, 2011; Holmes, 2012). Holmes (2006) has argued that three sets of forces: production; consumption; and protection, create seven distinct types of countryside in Australia, namely: productivist agriculture; rural amenity; pluri-active; peri-metropolitan; marginalized agriculture; conservation; and indigenous. Whilst indigenous people's use of land is very much a minority European interest (the Sami areas of Scandinavia might be an exception), the rest of Holmes' categories all resonate with types of region found in Europe.

Reflecting on Australia's immense rural territory, Holmes (2006, 2012) proposes a new model for understanding not only expectations for the rural, but also society's use of the rural. Holmes' model is grounded on multifunctionality as a key attribute of rural space, and he suggests that rural areas may be classified differently according to the relative importance of the modes of occupancy concerning production, consumption and protection. Their positioning helps us to understand what the 'vocation' of the areas is today. In this way, the conceptual model by Holmes (2006, 2012) is one basis in representing the different transition pathways rural areas might follow, not only centred on agriculture but including agriculture and its relations with other uses of, and demands on, the rural.

Further, Holmes also points out that the relative importance of the multiple functions of rural space is not constant and its relative importance may change (Holmes, 2012). These transition pathways need to be understood in order to identify the potential and possible developments of agriculture and define strategies for the farming sector.

In many rural regions of Europe, there are favourable biophysical and structural conditions for agriculture and the productivist paradigm has dominated for decades, with production the main driver of rural occupancy (Primdahl and Swaffield, 2010). However, social demand for non-commodity functions has typically been rising, related to environmental concerns, and also to expectations for recreation and life quality, leading to a new awareness concerning values of the physical landscapes and the need to consider other farming outcomes besides production.

On the other hand, in rural areas with limiting conditions for industrial agriculture, as in many places in southern Europe, agricultural systems have only recently (if at all) entered the productivist phase (Robinson, 2008; Perfecto *et al.*, 2010; Ortiz-Miranda *et al.*, 2013). The limitations for industrial agriculture are a function of a combination of many factors including natural conditions, location, structural constraints, lack of access to technology and socio-political history. In these areas, there is often a specific landscape character, with high interest for nature conservation. Such areas are thus often much valued by society due to conservation and consumption goals – but not well integrated in the modernization farming discourse and ideals (Pinto-Correia *et al.*, 2014). If the conservation and consumption roles could be acknowledged, this could lead to new forms of management and compensation, corresponding to new and emerging roles for farming otherwise in decay (van der Ploeg, 2008; Barbieri and Valdivia, 2010; Oreszczyn *et al.*,

2010). Notwithstanding the revitalization of marginal farming areas like Tuscany, with its privileged location near many cultural centres, its resurgent rural tourism and second home occupancy, there are also many parts of southern Europe, for example in parts of Spain, Portugal, Greece and Cyprus where 'desertification' proceeds apace and cultural landscapes contingent on high levels of manual labour (such as cultivation terraces) are falling into disarray and decay. The reversal of this decline, in the shadow of the growing threat of climate change, remains a major European challenge.

In between the extremes described above many possible combinations may exist, reflecting not only differences in the farming sector *per se*, but also differences in the present balance between production, consumption and conservation drivers (Holmes, 2012; van Eupen *et al.*, 2012) and the specificities of regional character in the different parts of Europe.

The typology of rural modes of occupancy proposed in this chapter aims to gauge the weight of the dimensions proposed by Holmes (2006, 2012) in different rural areas, so as to identify the likely trajectories of these areas. Figure 3.1 shows the balance between production, consumption and protection drivers in each of the six types of areas identified. Figure 3.2 shows the distribution of these types in Europe, at the NUTS 2 level.

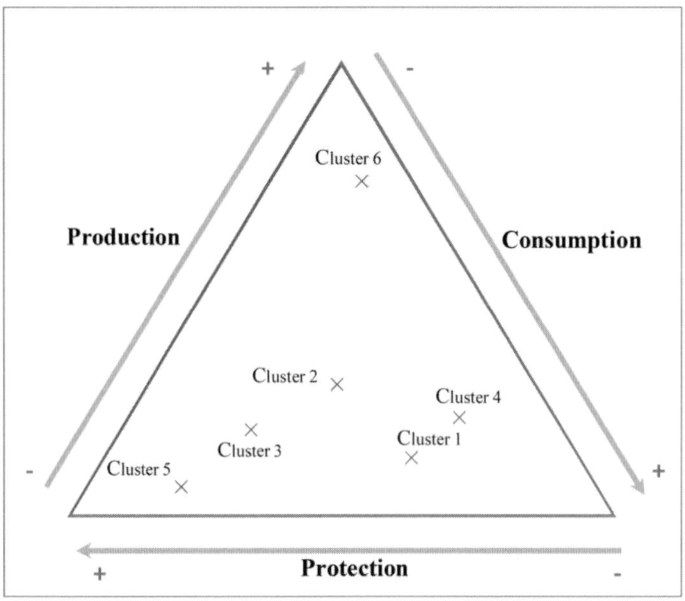

Fig. 3.1. The interplay between production, consumption and protection drivers in the rural space, in different types of regions in Europe. Here, the absolute role of each of these drivers is not reflected but rather their relative importance in relation to other drivers. Cluster 1: Consumption countryside; Cluster 2: Mixed countryside; Cluster 3: High nature value and consumption; Cluster 4: Consumption in the agricultural countryside; Cluster 5: High nature values; Cluster 6: Specialized agriculture (Source: authors).

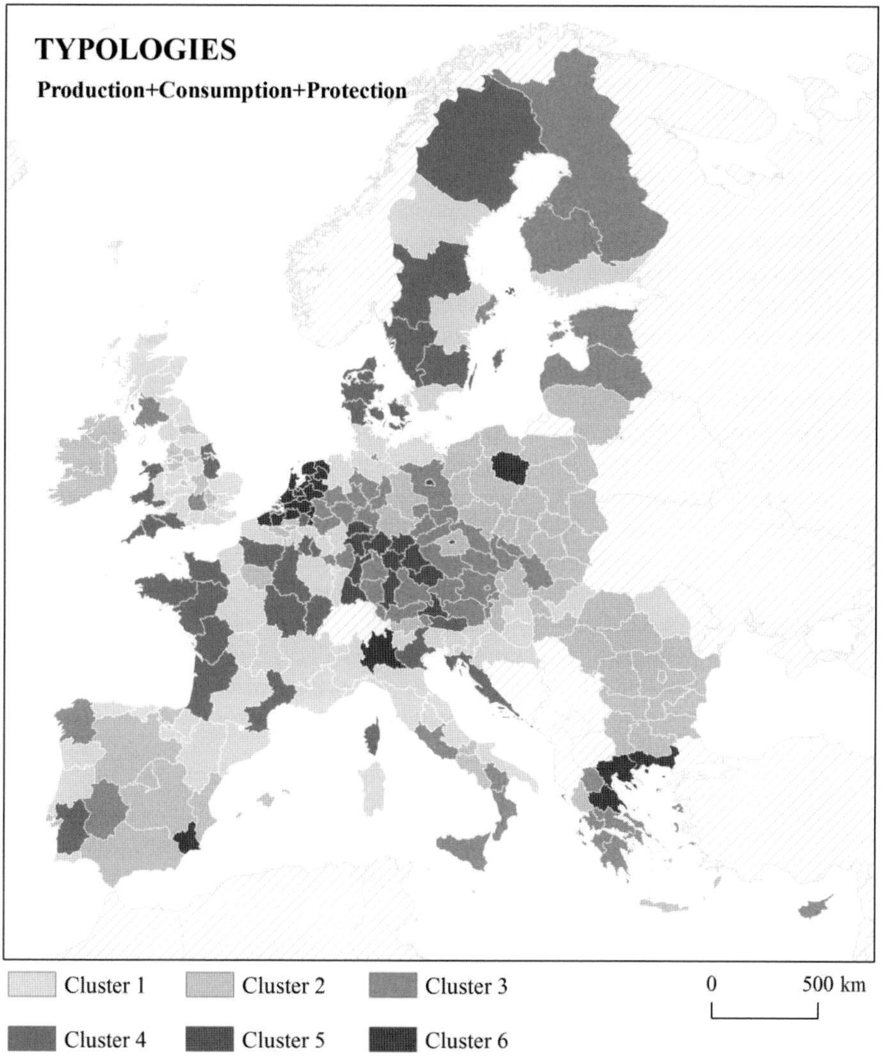

Fig. 3.2. Map of the distribution of types of regions in Europe according to their main modes of rural occupancy. For Cluster names, see Fig. 3.1 (Source: authors).

The aim of this classification exercise was to explore a combined typology of rural areas in Europe, where the balance between the role of production, consumption and protection as drivers of the use of rural space, is effectively assessed. Classification is not absolute but relative and is a tentative approach to how new modes of rural occupancy may be distributed in Europe. To obtain a reasonable level of detail, the NUTS 2 level has been selected as the scale of analysis. In order to identify each of the dimensions proposed by Holmes (2006, 2012), a few variables were selected. The selection of variables has been conditioned by the availability of data for the whole of Europe at the required scale. The intention was to identify variables which could express production intensity (share of total NUTS 2 area occupied by Utilized Agricultural Area; grazing livestock intensity; short rotation forests; employment rate in primary sector; rate of specialized farming), consumption pressures (nights spent in camping grounds; share of urban areas according to OECD urban/rural typology; UNESCO world heritage sites; employment in recreation activities) and protection interests (share of total NUTS 2 area occupied by Natura 2000 sites; share of rural areas according to OECD rural/urban typology; share of mountain areas; rate of holdings with organic farming), at the regional level NUTS 2.

The six clusters represent the combination of the three drivers of rural occupancy, and can therefore be characterized as: Cluster 1: Consumption countryside; Cluster 2: Mixed countryside; Cluster 3: High nature value and consumption; Cluster 4: Consumption in agricultural countryside; Cluster 5: High nature values; Cluster 6: Specialized agriculture.

As with any other typology, the picture would be changed if a few of the selected variables were excluded and other variables included. However, despite this limitation it is worthwhile reflecting on the challenges raised by this new combination of drivers affecting European rural areas. If we want to understand how agriculture will progress and how sustainable it may be in the future, we also need to understand the multiple drivers affecting rural areas which interact with the agricultural sector. In areas where agriculture has difficulty being competitive in global markets, the pathway towards sustainability in the future may be a different agriculture structure, sustained by those who consume the countryside such as lifestyle farmers and urban citizens. In areas with high conservation value, the pathway towards sustainability may be the acknowledgement and compensation of the conservation role of traditional farming. Such novel combinations are still to be fully acknowledged by the dominant administrative and policy regimes and therefore are not fully unfolded.

Conclusion

In this chapter we have noted the enormous diversity of land use and socio-economic conditions in rural Europe. Our understanding of change can be explored through the lens of individual land management units if our focus is household dynamics, but it became clear from the early 1990s that treating each farm as an independent economic entity failed to capture the empirical diversity of adjustment processes and the wider connections to regional economies. Farms were thoroughly embedded within distinct rural economies in many complex ways, particularly through off-farm pluriactivity, and increasingly, via a number of other means such as direct sales. Furthermore, farmers are embedded in local communities and these have been in a state of flux in many parts of Europe and are today diverse in terms of actors and networks with changing social values and new concerns, and are also often reshaping relations with urban communities. Consequently, understanding regional diversity has emerged as a subject of scrutiny and several authors have conceptualized new differences across rural Europe, in farming and in relation to the state of the rural as a whole.

Social science has been relatively quick to engage in these changing ruralities, once the baggage of old style rural sociology and narrowly conceived farm economics was left behind in favour of a broader view of rural economic space (Woods, 2011). Rural areas of Europe are now much more impacted by consumption of rural space than in the recent past but we should not forget that although there have been some significant changes in the policy architecture, support for production remains dominant in terms of the injection of public money; and if anything the productivist discourse has been strengthened by the prospect of what John Beddington has termed a 'perfect storm' of growing population, increased natural resource scarcity and climate change. New forms of regulation may have softened the harshest components of support for production, but a resurgent set of productivist values is now actively promoted under the guise of sustainable intensification.

There remain many parts of Europe where the idea of agricultural intensification is highly challenging, if not inconceivable. Here, there may be some scope for deconstructing core knowledge and adapting it to local conditions but that process of adaptation often hinges around new uses of rural spaces and activities that are less driven by ideas from the technological core and more from retro-innovation and new uses of rural space. It is rarely intensive in its impact on rural space and environments.

Behind the observable diversity, certain broad trends in rural land use can be discerned. There is much greater variety in the way that rural land is used and rural places have become more differentiated, but farming systems have often been simplified in core farming areas and enterprise size has grown. These areas often have strong agri-businesses and may be associated with intensification and environmental sustainability challenges. In non-core areas, different processes have driven change. These processes of differentiation have been forged both within the dynamics of the farm and in the regional context. Some farming areas have been heavily infiltrated by hobby farming; other areas are farmed extensively with a light touch and have been designated for their cultural landscapes and biodiversity. More marginal areas have tended to become even more marginal in terms of their contribution to total agricultural output, but as O'Connor *et al.* (2006) have shown, they can sometimes be transformed through what might be termed the 'Tuscan' model, where traditional agriculture can combine with new forms of consumption to create an alternative but nonetheless vibrant rural economy. We should perhaps be cautious of its applicability on a wide scale, though, for the preconditions for socio-economic revival in rural Tuscany may have been contingent on many factors not replicable elsewhere.

The Tuscan experience does, however, point towards a need to better understand innovation and this has been the subject of a great deal of work by Braczyk *et al.* (1998), Asheim *et al.* (2008) and others within regional economies and by Wiskerke and van der Ploeg (2004) in a specifically agrarian context. Van der Ploeg and Marsden (2008) move towards a more integrative approach and use the metaphor of 'unfolding webs' to describe new rural urban developments. It is clear that the old paradigms used to frame innovation opportunities need reframing. Innovation in the rural land use sector and the wider rural economy happens in different ways in different regions.

Since the early 1990s, in a post-Rio world, the issue of sustainability has become ever more salient, even if the economic crisis of the last decade has offered a distraction to many. Recently, in the outcome of the 2008 food crisis, new struggles for food sovereignty and regional autonomy have also created the conditions for the revival of rural communities and small-scale production associated with short supply chains.

The transition theory approach has endeavoured to capture this need for a transformation which is not just about business renewal but also about responding to global imperatives to develop a more sustainable economy. In recent years, the exploration of transitions towards more sustainable outcomes has given considerable attention to transition theory as developed by Dutch researchers (Kemp and Martens, 2007). Much of their early work explored large-scale industries rather than farming, but lately a number of projects have used the transition management lens to explore sustainability transitions in agriculture, including both SOLINSA and FarmPath. Agricultural heterogeneity throws up an enormous challenge to the implementation of a transition management approach. There is no dominant and distinctive regime of production which is normally the target of attention in transition studies. Farms and farmers are highly differentiated and probably becoming more so. But particular regions may be more or less dominated by a single regime. And it remains highly likely that regionally specific sustainability strategies for the farm and rural sector more generally, can be beneficially explored through the transition management lens. We follow up the emphasis on exploring new niches and their scope for up-scaling into new regimes in different parts of Europe in the core of this book.

References

Alanen, I. (2002) Soviet community spirit and the fight over the rural future of the Baltic countries. *Eastern European Countryside* 8, 15–29.

Arnalte-Alegre, E. and Ortiz-Miranda, D. (2013) The 'Southern Model' of European agriculture revisited: Continuities and dynamics. In: Ortiz-Miranda, D., Moragues-Faus, A. and Arnalte-Alegre, E. (eds) *Agriculture in Mediterranean Europe. Between Old and New Paradigms* (*Research in Rural Sociology and Development, Volume 19*). Emerald Group Publishing Ltd, Bingley, UK, pp. 37–74.

Asheim, B., Cooke, P. and Martin, R. (2008) Clusters and regional development: Critical reflections and explorations. *Economic Geography* 84, 109–112.

Barbieri, C. and Valdivia, C. (2010) Recreation and agro-forestry: Examining new dimensions of multifunctionality in family farms. *Journal of Rural Studies* 26, 465–473.

Beddington, J. (2009) Food, energy, water and the climate: A perfect storm of global events? Available from http://www.bis.gov.uk/assets/goscience/docs/p/

Braczyk, H-J., Cooke, P. and Heidenreich, M. (eds) (1998) *Regional Innovation Systems: The Role of Governances in a Globalized World*. Routledge, London, UK.

Bryden, J. (1988) *Farm Structures and Pluriactivity in Europe - Part One: Full Research Report.* ESRC End of Award Report, RD00232348. ESRC, Swindon, UK.

Chayanov, A.V. (1966) The theory of peasant economy. In: Thorner, D., Kerblay, B. and Smith, R.E.F. (eds) *The Theory of Peasant Economy*. Richard D. Irwin for the American Economic Association, Homewood, Illinois, pp. 29–269.

Copus, A. and Hörnström, L. (eds) (2011) *The New Rural Europe: Towards a Rural Cohesion Policy.* Nordregio report 2011:1. Available at: http://www.nordregio.se/en/Publications/Publications-2011

Curry, N. and Owen, S. (1996) Introduction, changing rural policy in Britain. In: Curry, N. and Owen, S. (eds) *Changing Rural Policy in Britain: Planning Administration, Agriculture and the Environment*. CCRU Press, Cheltenham, UK.

Darnhofer, I., Sutherland, L-A. and Pinto-Correia, T. (2015) Conceptual insights derived from case studies on 'emerging transitions' in farming. In: Sutherland, L-A., Darnhofer, I., Wilson, G.A. and Zagata, L. (eds) *Transition Pathways towards Sustainability in Agriculture: Case Studies from Europe*. CABI, Wallingford, UK, pp. 189–204.

de Groot, R. (ed.) (2010) Chapter 1. Integrating the ecological and economic dimensions in biodiversity and ecosystem services valuation. In: *The Economics of Ecosystems and Biodiversity: Ecological and Economic Foundations*. The Ecological and Economic Foundations, Brussels, Belgium, pp. 1–40.

Domon, G. (2011) Landscape as a resource: Consequences, challenges and opportunities for rural development. *Landscape and Urban Planning* 100, 338–340.

European Commission (2010) *A Resource-Efficient Europe – Flagship Initiative of the Europe 2020 Strategy*. European Commission, Brussels, Belgium.

Evans, N., Morris, C. and Winter, D.M. (2002) Conceptualizing agriculture: A critique of post-productivism as the new orthodoxy. *Progress in Human Geography* 26, 313–332.

Friedmann, H. (1980) Household production and the national economy: Concepts for the analysis of agrarian formations. *The Journal of Peasant Studies* 7, 158–184.

Gasson, R. and Errington, A. (1993) *The Farm Family Business*. CAB International, Wallingford, UK.

Gasson, R., Crow, G., Errington, A., Hutson, J., Marsden, T. and Winter, D.M. (1988) The farm as a family business: A review. *Journal of Agricultural Economics* 39, 1–41.

Goodman, D. and Redclift, M. (1985) Capitalism, petty commodity production and the farm enterprise. *Sociologia Ruralis* 25, 231–247.

Holmes, J. (2006) Impulses towards a multifunctional transition in rural Australia: Gaps in the research agenda. *Journal of Rural Studies* 22, 142–160.

Holmes, J. (2012) Cape York Peninsula, Australia: A frontier region undergoing a multifunctional transition with indigenous engagement. *Journal of Rural Studies* 28, 1–14.

Horlings, I. and Marsden, T.K. (2011) Towards a real green revolution: Exploring the conceptual dimensions of a new ecological modernization of agriculture that could feed the world. *Global Environmental Change* 21, 441–452.

Kemp, R. and Martens, P. (2007) Sustainable development: How to manage something that is subjective and never can be achieved? *Sustainability Science Practice and Policy* 3, 1–10.

Lobley, M., Errington, A. and McGeorge, A. with Millard, N. and Potter, C. (2002) *Implications of Changes in the Structure of Agricultural Businesses*. Final report for DEFRA, London, UK.

Marsden, T. (2013) From post-productionism to reflexive governance: Contested transitions in securing more sustainable food futures. *Journal of Rural Studies* 29, 123–134.

Marsden, T. and Sonnino, R. (2008) Rural development and the regional state: Denying multifunctional agriculture in the UK. *Journal of Rural Studies* 2, 422–431.

Marsden, T., Murdoch, J., Lowe, P., Munton, R. and Flynn, A. (1993) *Constructing the Countryside*. UCL Press, London, UK.

Marsden, T., Munton, R., Ward, N. and Whatmore, S. (1996) Agricultural geography and the political economy approach. *Economic Geography* 72, 361–376.

Mather, A.S., Hill, G. and Nijnik, M. (2006) Post-productivism and rural land use: Cul de sac or challenge for theorization? *Journal of Rural Studies* 22, 441–445.

Moragues-Faus, A., Ortiz-Miranda, D. and Marsden, T. (2013) Bridging Mediterranean agriculture into the theoretical debates. In: Ortiz-Miranda D., Moragues-Faus, A. and Arnalte-Alegre, E. (eds) *Agriculture in Mediterranean Europe. Between Old and New Paradigms* (*Research in Rural Sociology and Development, Volume 19*). Emerald Group Publishing Ltd, Bingley, UK, pp. 9–36.

Newby, H., Bell, C., Rose, D. and Saunders, P. (1978) *Property, Paternalism and Power: Class and Control in Rural England*. Hutchinson, London, UK.

O'Connor, D., Renting, H., Gorman, M. and Kinsella, J. (2006) The evolution of rural development in Europe and the role of EU Policy. In: *Driving Rural Development: Policy and Practice in Seven EU Countries*. Van Gorcum, Assen, the Netherlands.

Oreszczyn, S., Lane, A. and Carr, S. (2010) The role of networks of practices and webs of influences on farmers engagement with, and learning about, agricultural innovations. *Journal of Rural Studies* 26, 404–417.

Ortiz-Miranda, D., Moragues-Faus, A. and Arnalte-Alegre, E. (eds) (2013) *Agriculture in Mediterranean Europe. Between Old and New Paradigms* (*Research in Rural Sociology and Development, Volume 19*). Emerald Group Publishing Ltd, Bingley, UK.

Perfecto, I., Vandermeer, J. and Wright, A. (2010) *Nature's Matrix. Linking Agriculture, Conservation and Food Sovereignty*. Earthscan, London, UK.

Peters, G.H. (1991) Agriculture and the macro-economy: Presidential address. *Journal of Agricultural Economics* 42, 231–236.

Pinto-Correia, T. and Breman, B. (2009) The new roles of farming in a differentiated European countryside: Contribution to a typology of rural areas according to their multifunctionality. Application to Portugal. *Regional Environmental Change* 3, 143–152.

Pinto-Correia, T. and Kristensen, L. (2013) Linking research to practice: The landscape as the basis for integrating social and ecological perspectives of the rural. *Landscape and Urban Planning* 120, 248–256.

Pinto-Correia, T., Menezes, H. and Barroso, F. (2014) The landscape as an asset in Southern European fragile agricultural systems: Contrasts and contradictions in land managers attitudes and practices. *Landscape Research*, 39, 205–217.

Pinto-Correia, T., Gonzalez, C., Sutherland, L-A. and Peneva, M. (2015a) Lifestyle farming: Countryside consumption and transition towards new farming models. In: Sutherland, L-A., Darnhofer, I., Wilson, G.A. and Zagata, L. (eds) *Transition Pathways towards Sustainability in Agriculture: Case Studies from Europe*. CABI, Wallingford, UK, pp. 67–82.

Pinto-Correia, T., McKee, A. and Guimarães, H. (2015b) Transdisciplinarity in deriving transition pathways for agriculture. In: Sutherland, L-A., Darnhofer, I., Wilson, G.A. and Zagata, L. (eds) *Transition Pathways towards Sustainability in Agriculture: Case Studies from Europe*. CABI, Wallingford, UK, pp. 171–188.

Potter, C. and Tilzey, M. (2005) Agricultural policy discourses in the European post-Fordist transition: Neoliberalism, neomercantilism and multifunctionality. *Progress in Human Geography* 29, 581–600.

Primdahl, J. and Kristensen, L.S. (2011) The farmer as a landscape manager: Management roles and change patterns in a Danish region. *Geografisk Tidsskrift/ Danish Journal of Geography* 111, 107–116.

Primdahl, J. and Swaffield, S. (2010) (eds) *Globalisation and Agricultural Landscapes: Change Patterns and Policy Trends in Developed Countries*. Cambridge University Press, Cambridge, UK.

Robinson, G. (2008) Sustainable rural systems: An introduction. In: Robinson, G. (ed.) *Sustainable Rural Systems. Sustainable Agriculture and Rural Communities.* Ashgate, Farnham, UK, pp. 3–40.

Salaman, R. (1965) *The History and Social Influence of the Potato*. Cambridge University Press, Cambridge, UK.

Selman, P. (2009) Planning for landscape multifunctionality. *Sustainability: Science, Practice and Policy* 5, 45–52.

Short, C. (2008) Balancing nature conservation 'needs' and those of other land uses in a multifunctional context: High-value nature conservation sites in lowland England. In: Robinson, G. (ed.) *Sustainable Rural Systems. Sustainable Agriculture and Rural Communities*. Ashgate, Farnham, UK, pp. 125–144.

Sjoblom, S., Andersson, K., Marsden, T. and Skerratt, S. (2012) *Sustainability and Short-term Policies*. Ashgate Studies in Environmental Policy and Practice, Ashgate, Farnham, UK.

Tracy, M. (1989) *Government and Agriculture in Western Europe 1880-1988*, 3rd edn. Harvester Wheatsheaf, London, UK.

van Berkel, D.B. and Verburg, P.H. (2011) Sensitising rural policy: Assessing spatial variation in development options for Europe. *Land Use Policy* 28, 447–459.

van der Ploeg, J.D. (2008) *The New Peasantries. Struggles for Autonomy and Sustainability in an Era of Empire and Globalization*. Earthscan, London, UK.

van der Ploeg, J.D. and Long, A. (1994) *Born from Within: Practices and Perspectives of Endogenous Rural Development*. Van Gorcum, Assen, the Netherlands.

van der Ploeg, J.D. and Marsden, T. (2008) *Unfolding Webs*. Van Gorcum, Assen, the Netherlands.

van Eupen, M., Metzger, M.J., Pérez-Soba, M., Verburg, P.H., Van Doorn, A. and Bunce, R.G.H. (2012) A rural typology for strategic European policies. *Land Use Policy* 29, 473–482.

Wiggering, H., Ende, H-P., Knierim, A. and Pintar, M. (eds) (2010) *Innovations in European Rural Landscapes*. Springer, Berlin, Germany.

Wilson, G.A. (2001) From productivism to post-productivism … and back again? Exploring the (un)changed natural and mental landscapes of European agriculture. *Transactions of the Institute of British Geographers* 26, 77–102.

Wilson, G.A. (2007) *Multifunctional Agriculture: A Transition Theory Perspective*. CABI, Wallingford, UK.

Wiskerke, H. and van der Ploeg, J.D. (eds) (2004) *Seeds of Transition: Essays on Novelty Production, Niches and Regimes in Agriculture*. Van Gorcum, Assen, the Netherlands.

Woods, M. (2011) *Rural*. Routledge, London, UK.

Chapter 4

Utilizing the multi-level perspective in empirical field research: methodological considerations

P. Karanikolas[1], G. Vlahos[1], L-A. Sutherland[2]

[1]*Agricultural University of Athens (pkaranik@aua.gr);* [2]*James Hutton Institute, Aberdeen*

Introduction

In this chapter we identify the challenges of operationalizing the multi-level perspective (MLP) in empirical field research, and draw on experiences from field work in the FarmPath project as a case study of how these issues can be addressed. As described in Sutherland *et al.*a (this volume) the purpose of the FarmPath project – and the basis of this book – is to improve our understanding of transition processes in European agriculture. In particular, the research focuses on transitions in the making, emphasizing niches which are in the 'take-off' phase. The research also had a normative aim, specifically to understand sustainability transitions. Sustainability is understood as a process which takes into account regional differences in the forms and capabilities of the agricultural sector. In light of this definition, a transdisciplinary approach was adopted. Each of these orientations of the research poses specific challenges to the development of methodology.

In conceptualizing sustainability transitions, we utilize the multi-level perspective. The MLP is perhaps the most frequently used of a number of strands of research falling within the broader literature on socio-technical transitions. Other strands include strategic niche management (Caniëlsa and Romijnb, 2008; Schot and Geels, 2008) and transition management (Loorbach, 2007; Voß *et al.*, 2009), which share similar theoretical roots. The appeal of the MLP approach is in its ability to address the dynamics of large-scale socio-technical systems, and the sustainability challenges that they represent (Smith *et al.*, 2010). All three perspectives adopt the position that major change (or transition) occurs through interactions on three levels: niches (where innovations emerge and gain support); regimes (rules and established patterns of action); and landscapes (factors beyond the niche and regime levels). Over the past decade, an extensive literature has developed in which the MLP is applied in analyses of transition in various industries, including transportation, energy, public sanitation, food production and entertainment. However, in keeping with the roots of the MLP in evolutionary economics and the history of technology, these analyses predominantly focus on completed, historical transitions. The emphasis on understanding the transition from niche innovation to mainstream – which is conceptualized as occurring over the course of decades – requires an historical approach. In the research presented in this book, emphasis on transitions in the making at regional level necessitated the collection of empirical field research data, in addition to document analysis.

There is academic support for utilizing the MLP in the ex-post study of emerging transitions. One exception to the historical use of the MLP perspective is Elzen *et al.* (2011), whose recent work utilizing the MLP includes key informant interviews to complement document analysis in a study of socio-technical transition within the pork industry. Similar to the research presented in this book, Elzen *et al.*'s study focused on a transition in the making, integrating document analysis with empirical field research. However, Elzen *et al.*'s paper contains only a very brief description of methods. Conceptualizations of transition management and strategic niche management more commonly include field research. This reflects the research emphasis on steering ongoing transitions. In undertaking the research, scientists either undertake historical analysis of large scale transitions or focus-in on specific processes or actors in a combination of document review and field research (e.g. Späth and Rohracher, 2010; Budde *et al.*, 2012; Musiolik *et al.*, 2012). Even so, a very limited description of methods is usually provided by these authors in their research publications. As noted by Genus and Coles (2008), the lack of methodological description, particularly when using historical methods, introduces the potential for considerable analyst influence on the key points identified. We suggest that given the particular emphasis on sustainability transitions, specifically the intent for findings to inform active steering of transitions, more attention needs to be paid to the empirical and analytical methods underpinning the recommendations arising from these research processes.

The study of *emerging* transitions in different regional contexts poses a number of methodological challenges, in particular:

- How can we be certain that the observed phenomenon is not just an incremental change or a mere change in ordinary practices but rather the beginning of a radical transition?
- Where does the geographic specificity of regional characteristics fit within the MLP?
- How can the components of the three levels of the MLP model be distinguished in practice, particularly when they occur at different spatial scales (e.g. how can the area of overlap 'between' a niche and regime be assessed? Where does a niche occur, only at the 'micro' level, or within a regime as well?)
- How can emerging transition processes be compared across European countries and regions (as niches by definition are fluid and highly varied – how can a basis for comparison be established)?

Lack of conceptualization of geographic scale and space is another criticism of the MLP (Lawhon and Murphy, 2011). Besides the 'functional'/non-spatial character of basic concepts of the MLP, such as the regime, criticism centres on the fact that transitions are basically conceived as phenomena taking place along one dimension – time – whilst space is an exogenous background which does not actively contribute to the process of change in the socio-technical configuration. In addition, although a transition is by definition a radical change of a whole socio-technical configuration, the emphasis in the literature to date has been on transitions which have a strong 'technical' dimension. In particular, the niche is typically defined as a 'technological niche'. Although technology is clearly important to agricultural systems, recent research suggests that transition within agriculture also involves new organizational forms and practices. Can these new organizational forms constitute niches in their own rights?

There is, thus, considerable challenge in attempting empirical application of the MLP. All of the above mentioned issues pose some intriguing methodological challenges, as various concepts have both to be elaborated and operationalized, in order to enable field research. Nevertheless, methodological implications arise also from a series of other contributions of the MLP, such as its critique of the dominant paradigm of knowledge production and integration into a process of social change. This is a topic of major importance, since the MLP draws on a multitude of scientific disciplines, trying to integrate them in a coherent whole. The integrated production of knowledge necessitates both the active participation of various actors along with the involvement of knowledge derived from various scientific fields.

The aim of this chapter is to present some of the methodological challenges in the course of studying emerging transitions as well as the responses to these challenges within the FarmPath research project. In particular, the issue of inter- and trans-disciplinarity is raised in the second section, followed by some challenges in the use of core concepts of the MLP, such as the regime, the niche, the landscape and the interactions between a niche and a regime. In the fourth section, some challenges in the selection of case study countries and sites are explored, while the fifth section concerns the challenge of sustainability assessment in emerging transitions. The chapter concludes with an assessment of the challenges associated with the spatial character of an emerging transition.

What kind of research for an emerging transition?

One of the distinguishing characteristics of socio-technical transition studies in general, and the MLP in particular is interdisciplinarity: knowledge derived from divergent scientific fields. Recently, there has been a growing interest in integrating this diverse scientific knowledge with the experiences of non-experts (a range of stakeholders, who are engaged in a participatory process of social learning). Thus, transdisciplinarity occurs from the combination of interdisciplinarity and stakeholders' engagement (Grin, 2010). This integrated production of knowledge underpins both the definition of problems and the analysis for promoting sustainable development (Voß *et al.*, 2006). Also, Wiek *et al.* (2006) argue that the requirements of managing transitions (such as generation, integration and adaptation of knowledge and methods) should be fulfilled within a transdisciplinary setting involving scientists, experts and stakeholders.

These two methodological approaches permeate broader studies of transition. In the current study of emerging transitions, interdisciplinarity is manifest in several respects among team members, while trandisciplinarity appears in the interaction of researchers with a broad range of stakeholders. For both approaches, some problems have been encountered, such as different language, jargon, and different competencies and skills. Additionally, in the transdisciplinary approach a series of other problems have arisen concerning participants, such as differing priorities and objectives, as well as difficulties with recruitment, engagement and retention of interest. In any funded research process, the obligations of the researchers to the funding body inevitably constrain the freedom with which research objectives and activities can be co-developed with stakeholder groups.

Within the current research, some answers were given as a response to these challenges. In particular, the development of a *common glossary* was important. Scientists involved in the FarmPath project included economists, sociologists and human geographers. Development of a clear conceptual framework and glossary were essential to the translation of project objectives into field research. These terms also formed the foundation of the conceptual framework; definitions can, thus, be found in Darnhofer (this volume).

In order to engage stakeholders in the research process, National Stakeholder Partnership Groups (NSPGs) were formed in each study country, consisting of local stakeholders and end-users of the research effort. These groupings were identified at the outset in order to provide immediate input into the selection of case studies, with further members added later, in order to reflect on the specific case studies of initiatives and farming models chosen for each field research team. The NSPGs were fundamental to the research process, identifying initial visions for regional sustainability of agriculture at their first meetings, providing contextualization to the literature review on young people in agriculture, identifying initiatives and providing access to stakeholders for interviews, acting as the first and final participants in the regional scenario building exercises, identifying key policy and governance issues and assisting with the development of policy recommendations from research findings. Members of the NSPGs were able to maximize opportunities for dissemination of information to the broader stakeholder community throughout the project. Central to the NSPGs were representatives of local and regional stakeholder groups including farming organizations, retailers, processors, consumer groups, certification organizations and government agencies. The inclusion of micro, small and medium-sized enterprises was also encouraged in each NSPG.

Working with stakeholders requires specific skills in group facilitation and communication. To enable the effective participation of stakeholders in the research process, internal training was provided and external facilitators hired for key events, in order to support the transdisciplinary interaction. The research teams also engaged in reflexive assessment of their interactions through NSPG meetings, as well as collecting feedback on the meetings themselves, to ensure that the process was mutually beneficial.

Operationalizing the core concepts of the multi-level perspective

The need for an elaborated use of principles for "bounding and measuring niches, regimes and landscapes" which are then concretized into precise criteria has been vigorously argued in the literature (Smith *et al.*, 2010:446). Several reviews of the MLP perspective have noted that the basic concepts have not been consistently applied, or indeed consistently developed theoretically (Genus and Coles, 2008; Smith *et al.*, 2010). Inconsistent theoretical definitions of the regime, in particular, have been a source of criticism of the MLP (Markard and Truffer, 2008; Smith *et al.*, 2010). In historical analysis, the regime can be defined retrospectively based on the structures which appear to have been of greatest significance to the transition – albeit this reflects post-hoc rationalization of events. Depending on the nature of the function associated with the regime, some authors equate regime with sector (e.g. Raven, 2007 considers waste and energy regimes, equating them with the waste management and energy production sectors), whereas others, notably Geels (2002, 2004) emphasize that regimes are defined by rules, and utilize examples from within specific sectors (for example Geels's 2007 analysis of radio and recording regimes within the music industry). Smith *et al.* (2010) argue for the need to develop means of systematically defining and bounding the system under question, going so far as to suggest the definition of indicators and specific measures.

Thus, after the collection and analysis of sufficient background information for an emerging transition to be identified, *the definition of the incumbent regime* is the first issue to be addressed, as the very notion of transition is defined in relation to a regime transformation. This implies a description of the established socio-technical configuration which has already changed or is in the process of changing. In this present research, a regime is defined as "fairly stable rules, including cognitive routines, shared beliefs, capabilities and competencies, lifestyles and user practices, institutional arrangements and regulations, and legally binding contracts" (Darnhofer, this volume:19). The emphasis on rules is suggestive of function – that the regime fulfils an important societal function. As such, the criterion for the definition of a regime has been the fulfilment of a social need such as mobility, housing, food, energy, etc. Subsequently, within each regime a number of sub-regimes can be identified, following the logic of a system which consists of various interdependent sub-systems.

The challenge of defining regimes in relation to emerging transitions is the scale at which the regime is defined. As Smith *et al.* (2005) point out, what appears to be a radical transformation of a regime at one level may represent an incremental change to a larger regime, if the regime is defined at a higher level. For the purposes of consistency, and to reflect the emphasis of this book on agricultural systems, all 'thematic' chapters of this volume refer to initiatives that in one way or another relate to an agri-food regime. This agri-food regime contains the production, processing and marketing sub-regimes and, depending on the context, might develop some connections with other regimes such as the recreation regime. As such, the regime largely corresponds to the agricultural sector, although we recognize that the two are conceptually distinct.

A particular challenge to assessing transitions in the agricultural sector is the multiple functions of agriculture. Whereas by definition regimes serve a single societal function, it is well established in the literature that agriculture serves multiple functions such as amenity, environmental protection, and preservation of cultural heritage. These functions could be included as sub-regimes but fit poorly alongside processing, production and marketing sub-regimes all of which target production more broadly. In some cases, these multiple functions of agriculture have more in common with other sectors such as energy, tourism and housing. As such, we chose to identify the primary function of the agricultural regime as production of agricultural commodities, with environmental protection, energy production and recreational functions identified as distinct but overlapping regimes.

How do we analyse a regime? Regimes are not static entities; instead, they are perceived as dynamic configurations. In order to assess whether a transition is occurring, the *trends* that are embedded in a regime need to be identified. In the case of the agri-food regime, trends include: homogenization of production structures and commoditization of food consumer products; intensification and accompanying farm specialization and appropriation; enlargement and globalization of value chains; unequal distribution of value added and power across the agri-food system; and energy consumption. Accordingly, consideration of the '*counter-factual situation*' can be useful for identification of the dynamics (the evolution) of the dominant regime (in other words what would be occurring if a transition were not underway).

In terms of assessing the internal structure, one of the promising recent contributions in this respect is the detailed examination of a regime in terms of three dimensions: technical, human-societal and institutional, proposed by Elzen *et al.* (2012), which draws on a modified version of an analytical scheme initially proposed by Geels (2004). In this classification, all technology or production process-related aspects of a regime are included in the technical dimension. The human-societal dimension indicates all individual and collective actors as well as networks whilst institutions comprise the third dimension, divided into three distinct categories: cognitive or interpretative institutions; normative institutions; and economic institutions. Besides simplification, this seems to be a powerful and very useful analytical tool for two additional reasons. First, it can be used for a similar analysis of the niche(s), making clear both its internal structure and its comparability with the regime and, second, as a consequence it accommodates the exploration of the process of linking between the niche and the regime.

After the regime, the second 'level' in the MLP is the *socio-technical landscape*, conceived as a broad exogenous environment that as such is beyond the direct influence of regime and niche actors (Geels and Schot, 2010). In this research the landscape pressures and opportunities which create regime tensions were identified, such as: Common Agricultural Policy reforms; demographic decline; health crises; food safety concerns; increased environmental awareness and activism; consumption patterns; climate change; and agriculture as a producer of energy. The next step was to clarify the ways these pressures and opportunities impact on the regime and the niche and, more specifically, on the three dimensions identified above. Landscape pressures were also examined in relation to their contribution to the stabilization of the regime; the territorial level (European, national, regional, local) at which they caused changes in the policy framework; their degree of articulation towards a particular problem; and their role as supporting or hindering factors for the farm succession in the given regime.

Niches are the third basic component of the MLP framework. A niche is conceptualized as a protected space: a specific domain in which radical innovations can develop without being subject to the selection pressure of the prevailing regime (Kemp *et al.*, 1998). The description of the niche takes place along the same three dimensions which have already been used for the description of the regime: technical, human-societal and institutional. The competitive or symbiotic character of the relationship between the niche-innovations and the existing regime is also explored. A criterion which has been used for the distinction between a niche and the regime is the granting of various forms of subsidies in order to sustain the respective activity (Elzen *et al.*, 2012). However, several emerging transitions studied within this research involve grants to farmers (and other actors) for example in the context of multi-annual agri-environmental schemes, creating a need for a clarification about the character of the respective initiative: if those grants are considered as part of the protection environment, then we still have a 'niche', otherwise it is an integral part of a 'regime'.

A number of additional challenges, thus, emerge. As already noted, the MLP provides a theoretical apparatus which has been used in a historical perspective, for completed transitions. In the study of an emerging transition, caution is needed because of the possible implications concerning the *distinction between regime and socio-technical landscape elements*. In this respect, a telling example is product prices. In particular, when the notion of regime is used, as in 'global agri-food regime', then prices are endogenous since price formation takes place within the regime. However, the same term is used in these specific transition case studies, at the regional or even sub-regional level, in a different way. In these specific cases, prices are imposed by forces pertaining to the socio-technical landscape domain and can in no way be influenced by regime or niche actors. This is the case in energy, tourism and most of the agri-food regimes described in the case studies in this volume. However, the opposite is true in some agri-food regimes, for example when the product is not a commodity but a consumer good, or in agri-tourism and rural amenities sub-regimes. A further complication arises when there is no market, or existing market structures present important deficiencies, as in the case of public goods or quasi-public goods, where prices can be imposed externally, for example in the form of either agri-environmental compensations or as a result of open internal negotiations. Therefore, prices could be either an 'external' or 'internal' element of a regime. Another case of the same challenge is that relevant to policy measures, especially the CAP. It is important to distinguish between EU policy objectives and design, and locally implemented measures that have been mediated and influenced by national, regional or even local policy agendas, implying a continuum and not a binary distinction. As with any conceptualization, the MLP inevitably represents idealized distinctions which need to be negotiated and considered during operationalization.

The interactions between the niche(s) and the regime, thus, assume major importance to transition processes because they are most easily influenced by actor strategies. This means that the focal point of an emerging transition is the linking process between niche and regime, termed '*anchoring*' by Elzen *et al.* (2012). Anchoring denotes the way a novelty is linked ('anchored') with the technical, human-societal and institutional aspects: technological anchoring, network anchoring and institutional anchoring. An innovative action could take place in either a niche, a regime, or the overlapping niche-regime area, called a 'hybrid forum'. Therefore, the linking process refers to any of those three 'areas'. Hybrid actors are those regime actors who are engaged in innovative niche activities. Anchoring is an evolving process; hence there is no certainty about the final outcome of an initial development of niches. In each emerging transition, the factors that play the most important role in making the links between a niche and a regime more durable could be identified through a detailed exploration. Furthermore, as the niche-regime relationship is a continuous interactive process, after the analysis of 'anchoring' it is important to look again at the niche(s) and find out if anything has changed from its interaction with the regime.

Moreover, regime change is not simply a result of transferring practices from a niche to a regime. Every transition involves a 'translation' by actors; this translation could refer to sustainability problems, to adaptation of lessons or altering contexts (Smith, 2007). Which actors translate opportunities, where the process of translation takes place (whether in a niche, a regime or their overlapping area) and what kind of translation is involved, are all important issues in this respect.

Various learning processes are involved in niche-regime interactions. Along with the articulation of expectations and visions, and the building of social networks, Schot and Geels (2008) distinguish learning as one of the processes for successful development of a technological niche. In particular, learning processes refer to issues such as technical aspects and design specifications; market and user preferences; cultural and symbolic meaning; and regulations and government policy. Two forms of learning are usually distinguished: (i) single loop learning or first-order learning, which concerns new insights into the policy options (solutions) in the case of a given policy problem and a given policy context; and (ii) double loop learning or second-order learning, with new insights not only into the solutions to a certain problem but also into the problem itself and the context in which decisions take place (van de Kerkhof and Wieczorek, 2005). Thus, a learning process could result in a modification of the specific instruments that are used (first-order learning), but may also lead to change in the framework of ideas and standards within which actors customarily work, including their basic assumptions (second-order learning).

Finally, the role of policies in the transition has to be addressed, particularly in relation to strategies at different levels, regional and territorial development approaches, levels of policy-making and changes in the policy framework over time. Of particular importance are various policy measures as part of the incumbent regime, as well as policy changes in relation to the emergence of the niche(s) and the process of regime transformation.

Case study selection and research methods

The purpose of the FarmPath research was to identify findings that were of relevance at European level. This led to a methodological challenge in devising a procedure for the selection of countries in a way that would represent a wide variety of European contexts, particularly central and peripheral Europe, northern and southern Europe, and original and new entrant states. Research was, thus, undertaken in seven European states: Bulgaria, the Czech Republic, France, Germany, Greece, Portugal and the United Kingdom. The regions within each country were selected so as to reflect the heterogeneity of agriculture and the wide diversity of 'agri-food' models and initiatives such as organic farming, local food or non-food products and services which are of particular interest to the research (for example potential areas of best practice). In some regions, environmental aspects had been identified as the core challenge to sustainability, whereas in others the emphasis was on social aspects. Some of the initiatives and farming models include explicitly spatial elements, such as Less Favoured Areas (LFAs) or non LFAs, farming in Natura 2000 sites and/or supporting biodiversity conservation functions of the NATURA areas. The prevailing production system (intensive – extensive – very low input), as well as the existence of 'high nature value' farming systems in the areas under examination, also formed part of the criteria.

The resulting in-depth analysis is intended to identify the key societal actors (including technologies and institutions) in these initiatives, the role of these varying actors in transition processes, and how different institutional, socio-cultural, political and geographic contexts and access to technologies have resulted in different levels of success for these initiatives; this success would be measured in terms of social, economic and environmental sustainability. This analytical approach is used to clarify the social, institutional and technological innovation needs involved in scaling up and disseminating these initiatives.

Overall, primary data collection has been undertaken in 21 case studies within the participating countries, as described below.

Elaboration of the selection procedure – the evaluation matrix

The evaluation matrix is a set of criteria by which the candidate case studies were identified, described and subsequently selected. The purpose of the evaluation matrix was twofold. First, to propose a set of criteria for the selection of the case studies a multidimensional evaluation criteria matrix was developed. For a limited number of criteria (considered as essential) threshold values were set in order to decide which initiatives qualified for inclusion. Second, the evaluation matrix was utilized to classify case studies in order to provide a common basis for their selection and ensure comparability between countries. The final selection of cases is described in Table 4.1.

Table 4.1. The outcome of the clustering process (Source: authors).

Selected Cluster	Initiative
Renewable energy production	On-farm wind energy production (Aberdeenshire, UK) On-farm biogas production (Vysočina, Czech Republic) On-farm biogas production in Wendland-Elbetal Bio-energy region (Germany)
Lifestyle farming	Sustainable rural lifestyle (Zhelen, Bulgaria) New management in small-scale farming (Montemor-o-Novo, Portugal) Lifestyle land management (Aberdeenshire, UK)
Local quality and certification schemes	Integrating rural tourism and local food production for sustainable development (Elena, Bulgaria) A regional label for quality products and environmental protection (White Carpathians, Czech Republic) A local quality convention (Plastiras Lake, Greece)
Collaboration in agriculture	Collaborating for multifunctionality in the Montado silvo-pastoral system (Portugal) Formalized machinery and labour sharing (Scotland, UK) Citizen shareholder capital for regional value creation (Freiburg, Germany)
Alternative agri-food networks	New farmers' markets (Plzeňský region, Czech Republic) Short supply chains around the city of Rennes (France) Integration of local winemaking and conventional tourism in Santorini (Greece)
High nature value farming	New agricultural practices in protected areas (Bulgaria) Valuing the Mediterranean wild resources (Portugal) Landscape management in the St Amarin Valley (France)
Reducing the environmental impact of farming	Collective action to reduce green algae (Brittany, France) Groundwater protection through organic farming (Mangfall valley, Germany) Adaptation for survival: the case of peach producers in Imathia (Greece)

Note: *Each research cluster is given a chapter in this book, presenting findings from the case studies that were simultaneously conducted in three selected countries.*

In total, 59 initiatives were proposed for selection (between 7 and 10 in each of the 7 countries). These were identified by the NSPGs in conjunction with the scientists involved, on the basis that they represented initiatives and changes to agriculture that were important at regional level. Dossiers on potential case studies were compiled, and then the cases assessed in order to identify the most appropriate with regard to the following criteria:

- The temporal scale – initiatives that had started at least 5 years earlier;
- Regional impact – the initiatives should not be isolated to a single local area but exist in other regions or different countries;
- Stage of transition – having moved beyond the stage of niche formation, as evidenced by the range and complexity of the actors involved; and
- Sustainability – initiatives should make a contribution to economic, social and environmental sustainability.

Once the cases were classified, a set of additional qualifications were taken into account for the selected initiatives: first, the inclusion of some failed initiatives (negative influence on sustainability) (such as cases where although the circumstances had been favourable, the transition process did not occur); second, the inclusion of some cases of completed transitions, as well as transition processes of both top-down and bottom-up character; third, two criteria for selection within clusters, that is the range of transition progress and good fit or comparability; fourth, inclusion of at least one mainstream and one alternative initiative, selected by each national team.

Research methods

In terms of research methods, the research was conducted in the form of qualitative case studies, utilizing a combination of qualitative interviews, focus groups and document review. The number of interviews and documents reviewed varied between the case studies, reflecting the availability of data. In most cases, approximately 10 to 15 interviews were undertaken with a range of stakeholders and participants. In some cases, however there was considerable documentation available for review and so fewer interviews were undertaken (e.g. biogas production in Wendland-Elbetal). Several of the cases also chose to undertake focus groups with participants (such as in the research on formalized machinery and labour sharing in the UK). Qualitative approaches are appropriate for the assessment of transitions in progress, as they are particularly well suited to exploring complex issues, and for studying processes that occur over time (Ritchie and Lewis, 2008). The flexible nature of qualitative research also enables the researcher to 'follow' the data; to identify new issues for exploration, and identify associated key informants. Assessments utilizing the MLP also increasingly integrate other theoretical perspectives to address specific aspects (e.g. Elzen *et al.*, 2011 who integrate social movement theory and political science with the MLP). This approach was also undertaken in some of the case studies, in order to further develop findings. For example, Diaz *et al.* (2013) combined the MLP with actor-network theory, while Sutherland and Holstead (2014) use the theory of planned behaviour to explore the role of renewable energy production in farm business decision-making. For consistency purposes, the analysis presented in this book is solely based on the MLP.

The challenge of sustainability assessment

As part of the research process, researchers conducted sustainability assessments of the case studies. This was necessary to assess whether the transition processes identified were leading towards greater sustainability in agriculture. Within the project as a whole, sustainability was identified as a process rather than an end result and one that is best assessed at regional level, reflecting the resources and needs of the region in question. Regional sustainability of agriculture was, thus, conceptualized as a quality of the regional farming system that emerges from processes of adaptation by members of the agricultural production and consumption network who respond to the changing needs and preferences of consumers and citizens. Sustainability assessments were conducted at regional level, in collaboration with the NSPGs.

Within the transition literature, sustainability assessment (SA) is perceived as a participatory, cyclical and transformative process, contributing to a clear understanding of context-specific sustainability problems and to the identification of desired ways of achieving solutions. The technique of *integrated sustainability assessment* also plays an important role in transition theories, as a constituent part of the integration of different research disciplines with each other and with stakeholders' knowledge (Weaver and Rotmans, 2006). In such a participatory process, social learning and non-linear knowledge production are the focal points, while sustainability is perceived as a normative orientation (Rotmans and Loorbach, 2010). Through the active involvement of stakeholders, SA aims to achieve:

- A shared perception of sustainability problems;
- The development of a vision for a sustainable future;
- The pursuit of possible solutions; and
- Learning about key relationships and ways of reframing problems and solutions.

By definition, a transition is about structural change of a regime to a more sustainable future. The need for a more concrete account of sustainability assessment is seen in different aspects of an emerging transition study, such as how a niche is more sustainable than the incumbent regime; how the various scenarios are assessed in terms of sustainability; and what learning aspects emerge from the SA of a transition.

European countries, regions and initiatives usually have different sustainability priorities. This variability calls for flexibility to be systematically incorporated into the assessment process. A common methodology must be adaptable to incorporate the wide range of economic, social, environmental, administrative and cultural circumstances found across Europe (Knickel *et al.*, 2009) and allow for a context-dependent selection and/or adaptation of indicators.

However, there is still a high level of criticism over the attempt to create an 'objective' sustainability assessment using elaborate indicator systems. The meticulous nature of the process tends to obscure the fact that any sustainability assessment is a series of trade-offs from both within and across the sustainability dimensions. It is the society, and not the experts, who should decide on the trade-offs. The latter approach, however, attributes more weight to the issue of representation in participatory procedures.

Based on the above arguments, a procedure for SA in emerging transitions was developed. Through a consultation process between researchers and stakeholders, the main sustainability issues of the region were identified and classified according to their relation to the three sustainability pillars (environmental, economic, social). These sustainability issues were then connected with the niche dimensions and niche-regime interactions. More specifically, these sustainability issues were used as one axis of a matrix, whilst the 'dimensions' of primary activities and impacts of the initiative were placed on the other axis. Three such matrices were constructed, one for each pillar of sustainability. These inductively identified 'sustainability criteria' were then compared with lists of indicators proposed by the research team, based on established indicator sets, such as the 'Signals' series of reports by the European Environmental Agency (2002). Using the indicator list, appropriate indicators were selected for the issue and dimension concerned. There could be more than one indicator per cell: the same indicator could be used for more than one issue. The impact of the initiative on a specific indicator was graded by a (+) positive, (-) negative or (=) invariant. This assessment was accompanied by a justification and (where possible) quantification of the grading. The change in an indicator towards one direction could be positive for one sustainability pillar or issue, while negative for another. A good example of this is the expansion of an irrigated area: it is a positive development in terms of economic sustainability, while negative in terms of environmental sustainability.

An 'a-spatial' transition?

To date, it has been mainly the temporal aspects of transitions that have been explored through the MLP, with the core concept of the 'regime' defined in terms of a functional space. However, recent research on sustainability has shown the lack of a clear conceptualization of the geographical dimension of historical and emerging transition processes (Cooke, 2010; Coenen and Truffer, 2012).

In addition, other strands of literature suggest the pursuit of sustainability through the *relocalization* of modern agriculture and food systems (Feagan, 2007). For example, efforts to analyse contemporary agriculture from a transition-to-sustainability perspective involve a serious critique of the 'modernization' paradigm in agriculture; this implies, inter alia, the reconnection of agriculture with specific agro-ecological contexts (Roep and Wiskerke, 2004). Moreover, agriculture has always been a highly heterogeneous activity. Various structural characteristics are always intertwined with sectoral specialization within concrete geographical and territorial contexts. The spatial character of recent restructuring developments in the global agri-food system have also been identified, despite forceful concentrating trends. This is evident, as "models of emerging alternatives can help relocalize production/consumption relationships in the food system in equitable ways. In other words, in relationships that are personalised and sustainable, and that are embedded in place and community" (Hendrickson and Heffernan, 2002:361). The spatial embeddedness of the agri-food system is also stressed recently by Marsden (2013), along with the need for the creation of policy spaces for more place-based forms of reflexive governance in the study of transitions to sustainability.

In the light of these findings, the focus of this research on *regional* scale and the *agri-food* sector, has some methodological implications, related to the spatial character of the transitions under study. The emphasis on the regional scale permeates the analytical framework of this research (see Darnhofer, this volume). In addition to this overall emphasis on the regional scale, the current research includes various explicit references to the regional dimension/character of the transitions. These include mainly contextual information on each case study concerning, for example, size, structural characteristics, typologies of rural areas, and extending up to various institutional arrangements and networks of actors.

The analysis of the material from the current research would benefit from a more explicit spatial account of transitions to sustainability, as the above mentioned findings from the literature suggest. In analytical and methodological terms this would imply clarifying the spatial variety in regime structures and landscape forces (Coenen *et al.*, 2012) as well as understanding how the functional socio-technical spaces of niche, regime and landscape relate to other dimensions of space, such as territorial, administrative and communicative spaces (and their particular topologies) (Smith *et al.*, 2010), a theme that has been prioritized for future research in sustainability studies. In a similar vein, the study of emerging transitions has to address the differentiation of regime, niche, and innovation system structures in specific regions of the world (Markard *et al.*, 2012) and analyse transitions as interdependent processes between territorialized, local and trans-local networks within the context of (changing) multi-scalar, institutional structures (Coenen *et al.*, 2012).

Conclusion

This research focuses on some novel aspects of transition studies, particularly emerging transitions occurring in the agri-food sector at the regional scale (see Darnhofer, this volume). Intriguing methodological challenges stem from the application of the MLP to emerging transitions. Responses to these challenges include strategies that were implemented in order to cope with interdisciplinarity problems as well as the active involvement of stakeholders in the context of NSPGs, in a transdisciplinary perspective.

This research has been conducted with some difficulties due to the elusiveness of the ideas which form the core of the research question (transition towards regional sustainability). In an effort to overcome these difficulties, various analytical endeavours have been applied. These are an alternative definition of the agri-food regime and its constituent parts on the basis of the multiple functions of agriculture; the detailed examination of both a niche and a regime in terms of technical, human-societal and institutional dimensions; and the linking process between niche and regime along the same three dimensions ('anchoring'). The focus on the regional scale in the research of an emerging transition has some noteworthy implications, especially in relation to the characterization of some factors as 'internal' or 'external' to the regime. This was exemplified by the cases of prices and policy measures.

The lack of relevant literature, particularly pertinent to the temporal and spatial levels this research is aiming at, as well as the need for an operationalization of the theoretical concepts, enabled the researchers to establish an evaluation matrix comprising 'transition' thresholds. This provided the criteria and the degree of fulfilment that were used in order to distinguish a 'potential transition' process from a mere practice change or a farm level 'trajectory'.

Furthermore, a procedure for sustainability assessment in emerging transitions has been proposed, involving a consultation process between researchers and stakeholders for the identification of the main sustainability issues of the region concerned, their classification according to their relation to the three sustainability pillars (environmental, economic, social), the selection of relevant indicators and the impact of the initiative on the specific indicator.

Given the re-localization trends in contemporary agri-food systems, the inherent heterogeneity of agriculture and the focus of this research on the regional sustainability of agriculture, further analysis should take a closer look at the issue of space-specificity in the study of transitions to sustainability, most notably the spatial variety in regime structures and landscape forces as well as the territorialized character of institutional embeddedness.

Of course, many issues remain open. However, the operationalization of basic concepts and their consistent theoretical development could make a substantial contribution not only to the empirical research through the identification of specific criteria but also to the advancement of transition theories as well.

References

Budde, B., Alkemade, F. and Weber, K.M. (2012) Expectations as a key to understanding actor strategies in the field of fuel cell and hydrogen vehicles. *Technological Forecasting & Social Change* 79, 1072–1083.

Caniëlsa, M.C.J. and Romijnb, H.A. (2008) Strategic niche management: Towards a policy tool for sustainable development. *Technology Analysis & Strategic Management* 20, 245–266.

Coenen, L. and Truffer, B. (2012) Places and spaces of sustainability transitions: Geographical contributions to an emerging research and policy field. *European Planning Studies* 20, 367-374.

Coenen, L., Benneworth, P. and Truffer, B. (2012) Towards a spatial perspective on sustainability transitions. *Research Policy* 41, 968–979.

Cooke, P. (2010) Regional innovation systems: Development opportunities from the 'green turn'. *Technology Analysis & Strategic Management* 22, 831–844.

Darnhofer, I. (2015) Socio-technical transitions in farming. Key concepts. In: Sutherland, L-A., Darnhofer, I., Wilson, G.A. and Zagata, L. (eds) *Transition Pathways towards Sustainability in Agriculture: Case Studies from Europe*. CABI, Wallingford, UK, pp. 17–32.

Diaz, M., Darnhofer, I., Darrot, C. and Beuret, J-E. (2013) Green tides in Brittany: What can we learn about niche-regime interactions? *Environmental Innovation and Societal Transitions* 8, 62–75.

Elzen, B., Geels, F.W., Leeuwis, C. and van Mierlo, B. (2011) Normative contestation in transitions 'in the making': Animal welfare concerns and system innovation in pig husbandry. *Research Policy* 40, 263–275.

Elzen, B., van Mierlo, B. and Leeuwis, C. (2012) Anchoring of innovations. Assessing Dutch efforts to harvest energy from glasshouses. *Environmental Innovation and Societal Transitions* 5, 1–18.

European Environmental Agency (2002) *Environmental Signals*. European Environment Agency, Copenhagen, Denmark.

Feagan, R. (2007) The place of food: Mapping out the 'local' in local food systems. *Progress in Human Geography* 31, 23–42.

Geels, F.W. (2002) Technological transitions as evolutionary reconfiguration processes: A multi-level perspective and a case-study. *Research Policy* 31, 1257–1274.

Geels, F.W. (2004) From sectoral systems of innovation to socio-technical systems. Insights about dynamics and change from sociology and institutional theory. *Research Policy* 33, 897–920.

Geels, F.W. (2007) Analysing the breakthrough of rock 'n' roll (1930–1970). Multi-regime interaction and reconfiguration in the multi-level perspective. *Technological Forecasting & Social Change* 74, 1411–1431.

Geels, F.W. and Schot, J. (2010) The dynamics of transitions: A socio-technical perspective. In: Grin, J., Rotmans, J. and Schot, J. (eds) *Transitions to Sustainable Development: New Directions in the Study of Long Term Transformative Change*. Routledge, New York, pp. 11–104.

Genus, A. and Coles, A.M. (2008) Rethinking the multi-level perspective of technological transitions. *Research Policy* 37, 1436–1445.

Grin, J. (2010) Understanding transitions from a governance perspective. In: Grin, J., Rotmans, J. and Schot, J. *Transitions to Sustainable Development: New Directions in the Study of Long Term Transformative Change.* Routledge, New York.

Hendrickson, M.K. and Heffernan, W. (2002) Opening spaces through relocalization: Locating potential resistance in the weaknesses of the global food system. *Sociologia Ruralis* 42, 347–369.

Kemp, R., Schot, J. and Hoogma, R. (1998) Regime shifts to sustainability through process of niche formation: The approach of strategic niche management. *Technology Analysis and Strategic Management* 10, 175–196.

Knickel, K., Kröger, M., Bruckmeier, K. and Engwall, Y. (2009) The challenge of evaluating policies for promoting the multifunctionality of agriculture: When 'good' questions cannot be addressed quantitatively and quantitative answers are not that good. *Journal of Environmental Policy and Planning* 11, 347–367.

Lawhon, M. and Murphy, J.T. (2011) Socio-technical regimes and sustainability transitions: Insights from political ecology. *Progress in Human Geography* 36, 354–378.

Loorbach, D. (2007) *Transition Management: New Mode of Governance for Sustainable Development.* Erasmus Universitiet. International Books, Utrecht, the Netherlands.

Markard, J. and Truffer, B. (2008) Technological innovation systems and the multi-level perspective: Towards an integrated framework. *Research Policy* 37, 596–615.

Markard, J., Raven, R. and Truffer, B. (2012) Sustainability transitions: An emerging field of research and its prospects. *Research Policy* 41, 955–967.

Marsden, T. (2013) From post-productionism to reflexive governance: Contested transitions in securing more sustainable food futures. *Journal of Rural Studies* 29, 123–134.

Musiolik, J., Markard, J. and Hekkert, M. (2012) Networks and network resources in technological innovation systems: Towards a conceptual framework for system building. *Technological Forecasting & Social Change* 79, 1032–1048.

Raven, R. (2007) Co-evolution of waste and electricity regimes: Multi-regime dynamics in the Netherlands (1969-2003). *Energy Policy* 35, 2197–2208.

Ritchie, J. and Lewis, J. (2008) *Qualitative Research Practice.* Sage, London, UK.

Roep, D. and Wiskerke, J.S.C. (2004) Reflecting on novelty production and niche management in agriculture. In: Wiskerke, J.S.C. and van der Ploeg, J.D. (eds) *Seeds of Transition, Essays on Novelty Production, Niches and Regimes in Agriculture.* Van Gorcum, Assen, the Netherlands, pp. 341–356.

Rotmans, J. and Loorbach, D. (2010) Towards a better understanding of transitions and their governance: A systemic and reflexive approach. In: Grin, J., Rotmans, J. and Schot, J. (eds) *Transitions to Sustainable Development: New Directions in the Study of Long Term Transformative Change.* Routledge, New York, pp. 105–222.

Schot, J. and Geels, F. (2008) Strategic niche management and sustainable innovation journeys: Theory, findings, research agenda, and policy. *Technology Analysis & Strategic Management* 20, 537–554.

Smith, A. (2007) Translating sustainabilities between green niches and socio-technical regimes. *Technology Analysis & Strategic Management* 19, 427–450.

Smith, A., Stirling, A. and Berkhout, F. (2005) The governance of sustainable socio-technical transitions. *Research Policy* 34, 1491–1510.

Smith, A., Voβ, J-P. and Grin, J. (2010) Innovation studies and sustainability transitions: The allure of the multi-level perspective and its challenges. *Research Policy* 39, 435–448.

Späth, P. and Rohracher, H. (2010) 'Energy regions': The transformative power of regional discourses on socio-technical futures. *Research Policy* 39, 449–458.

Sutherland, L-A. and Holstead, K. (2014) Future-proofing the farm: On-farm wind turbine development in farm business decision-making. *Land Use Policy* 36, 102–112.

Sutherland, L-A., Wilson, G. A. and Zagata, L. (2015a) Introduction. In: Sutherland, L-A., Darnhofer, I., Wilson, G.A. and Zagata, L. (eds) *Transition Pathways towards Sustainability in Agriculture: Case Studies from Europe.* CABI, Wallingford, UK, pp. 1–16.

van de Kerkhof, M. and Wieczorek, A. (2005) Learning and stakeholder participation in transition processes towards sustainability: Methodological considerations. *Technological Forecasting & Social Change* 72, 733–747.

Voß, J-P., Bauknecht, D. and Kemp, R. (2006) *Reflexive Governance for Sustainable Development*. Edward Elgar Publishing, Cheltenham, UK.

Voß, J-P., Smith, A. and Grin, J. (2009) Designing long-term policy: Rethinking transition management. *Policy Sciences* 42, 275–302.

Weaver, P. and Rotmans, J. (2006) Integrated Sustainability Assessment: What is it, why do it and how. *International Journal of Innovation and Sustainable Development* 1, 284–303.

Wiek, A., Binder, C. and Scholz, R.W. (2006) Functions of scenarios in transition processes. *Futures* 38, 740–766.

Chapter 5

Lifestyle farming: countryside consumption and transition towards new farming models

T. Pinto-Correia[1], C. Gonzalez[1], L-A.Sutherland[2], M. Peneva[3]

[1]*University of Évora (mtpc@uevora.pt);* [2]*James Hutton Institute, Aberdeen;* [3]*University of National and World Economy, Sofia*

Introduction

This chapter deals with the role of 'countryside consumption' as a driver of change in agricultural and land-based activities in Europe. It analyses how increasing utilization of the countryside as a space of consumption (such as for amenity, living space, and leisure activities) has led to a shift in the role of production in relation to other land management activities, particularly in areas where production has become marginal. The chapter also assesses how countryside consumption manifests in the self-concept of farmers and the agricultural sector, how practical and social changes resulting from the transition inherent in increased levels of lifestyle farming are acknowledged by the agricultural sector, and how the agricultural sector, in this context, interacts with other players in rural areas.

The concept of countryside consumption has been constructed as part of a transition to a differentiated countryside, for instance in England, whereby productivist industry operates alongside post-productivist activities such as nature conservation and amenity (Murdoch *et al.*, 2003). In his Australian research, John Holmes (2006) introduced the concept of functional trajectories in rural areas where the relative weight of production, consumption and protection functions are being altered, as consumption and protection contest the former dominance of production values. These changes in the relative weight of the three main functions result in new and complex modes of rural occupancy (e.g. new modes of human use of the rural space) (Holmes, 2006, 2012; van der Ploeg and Marsden, 2008; van Berkel and Verburg, 2011).

In the transitional trajectories defined by van der Ploeg (2009), consumption is currently associated with 'deactivation', characterized by a reduction of agricultural activities in rural areas, and a shift towards leisure, nature reserves, rural dwellings, and bioenergy production. These are most often the result of contradicting and complex dynamics formed by the interaction of diverging processes. These processes include 'industrialization', which is less relevant in the areas where consumption is taking the lead. There is also 're-peasantization', characterized by the active construction of new degrees of autonomy on commodity markets (in other words major farm resources are produced within the farm itself) and the presence of new actors in farming. Re-peasantization is shown to be highly relevant in areas where consumption trends increase. Along the lines of these trends,

 Transition pathways towards sustainability in agriculture: Case studies from Europe (eds L-A. Sutherland *et al.*)

Marsden (2013) recently identified the emergence of a renewed interest in production functions in farming areas in Western countries, particularly after the food crisis of 2007/2008, suggesting a further shift taking place from post-productivism to sustainable intensification (bio-economic productivism).

Countryside consumption can thus be understood as a driver of farm and farmland management grounded in quests for a rural lifestyle, and healthy food and leisure, which may or may not be closely linked to production. Countryside consumption definitively changes the way in which the actors concerned deal with farming, or involves the introduction of new actors into the farming sector. This particular form of farm management can be defined as 'lifestyle farming', where the rural landholder does not derive his/her income primarily from production (in other words the income generated from agriculture is not the main driver of land use and the value of agricultural production tends to be less of a determinant than other factors for farmer choices). Yet the lifestyle farmer may be, and often is, a producer and since he/she manages agricultural land, countryside consumption also has an impact on the management of the physical landscape.

In this analysis, lifestyle farming is conceptualized as forming a socio-technical niche (see Darnhofer, this volume), since it introduces novel land uses. The concept relates to new beliefs and values, new technologies and practices, new configurations of actor groups, new networks, and it may lead to new policies, or to a renewed use of the existing policy framework. In essence, lifestyle farming represents a mismatch with existing commercial farming structures and practices (the 'regime') and it also deviates from the emerging push towards sustainable intensification. Although the shift towards lifestyle farming has been growing for several decades, most often there has been no collective or shared intention to enable radical change at the regime level; it is a change that originated among local actors, taking place at local levels in various locations throughout Europe. However, niche actors or groups of actors engage with organizational structures at regime level in increasingly significant ways. Here in particular, because of its fuzzy character, relations with relevant regimes including housing and conservation are complex and require a detailed analysis within an updated conceptual framework, in order to be disentangled. Lifestyle farming thus presents an opportunity to further develop and challenge the concepts within the multi-level perspective (MLP) in assessing change processes surrounding transition in the agricultural sector. In turn, the complexity of the processes of countryside consumption and its multiple dimensions require new analytical tools and conceptual grids from, among others, the agronomy and rural community perspectives (Renting *et al.*, 2009; Milone and Ventura, 2010). Using the MLP as an analytical framework therefore provides us with a tool to achieve a more comprehensive understanding of these processes.

Characterizing lifestyle farming as a transition in progress

The case studies

In order to assess the transition to lifestyle farming evident in specific European regions, three case studies were selected (Fig. 5.1). The Bulgarian case focuses on the Trinoga Association, included in a formally organized initiative which, since 2005, has promoted the idea of community-supported agriculture for healthy and locally grown food. The association is located in the village of Zhelen, situated 50 km north of Sofia, a depopulated mountain area with small-scale farming. The initiators are young people with higher education and urban backgrounds who settle in the village, producing their own food and developing new activities of public benefit for the local community. Farm structure is determined by the terrain and natural conditions. The ownership of the land is fragmented, as the average size of the farms is 0.4 ha. The British initiative concerns lifestyle farmers in Aberdeenshire, Scotland: households living on and managing land holdings of less than 10 ha for recreational and life quality purposes. Although small-scale self-provisioning has a lengthy heritage in Scotland, dating back to the 18th and 19th centuries, the initiative described relates to recreational small-scale land use which evolved primarily since the 1970s, with the arrival of the oil industry (and associated wealth) in Aberdeen. This occupation of agricultural land for lifestyle purposes experienced a boom from 2003 to 2008, but was negatively impacted by the post 2008 recession. Finally, the Portuguese study also reflects a spontaneous, non-organized process: rural small farms (from 2 to 20 ha) in the area surrounding Montemor-o-Novo in the Alentejo region, are increasingly occupied by lifestyle farmers with an urban background. Montemor-o-Novo has a beautiful landscape and is located 100 km east of Lisbon, at the axis between Lisbon and Madrid. The replacement of former local inhabitants and farm families by newcomers started in the late 1980s, continued into the 1990s, and has clearly been increasing in the last 10 years. The Aberdeenshire and Montemor-o-Novo cases thus represent on-going processes, whereas the Zhelen case represents a formally organized initiative.

The temporal scale of the three initiatives is not the same, with the Scottish case dating from the 1970s, the Portuguese from the late 1990s, and the Bulgarian from only the past 5 years. The spatial scale also varies: a small village in Bulgaria, a surrounding area to a small town in Portugal, and the whole of Aberdeenshire in Scotland. The studied processes are thus in different phases, with correspondingly different scaled impacts at regional level. Nevertheless, in all three cases there is an expansion of lifestyle farming, both in the number of farms and in the acknowledgement of this new type of land management by local communities. The drives associated to this new type of farming range from life quality to proximity to nature and to a search for healthier food, differing from those of conventional commercial farmers.

Lifestyle farming as a transition in progress

The basis of the MLP is the idea that systems (regimes) are 'locked-in' to a steady trajectory. Innovations emerge inside and outside of regimes, requiring development and protection from mainstream regime market trends by niche actors. These niches can lead to regime change when a shift occurs in the external context (i.e. landscape), which destabilizes the regime and opens a 'window of opportunity' for niche expansion, typically in the form of an increased demand for the technology lodged within the niche (for further details, see Darnhofer, this volume). In this section we explore the development of lifestyle farming in the three study sites utilizing MLP.

Niche development

As discussed by Darnhofer in Chapter 2 of this volume, niches are conceptualized as being created and protected by small groups of actors, working on innovations that deviate from existing regimes. These actors develop norms and expectations which are less stable than those for regime actors but nevertheless guide and constrain the development of the niche. The socio-technical system – the technologies involved in the niche as well the material and social context for action – also constrain and enable development in specific directions.

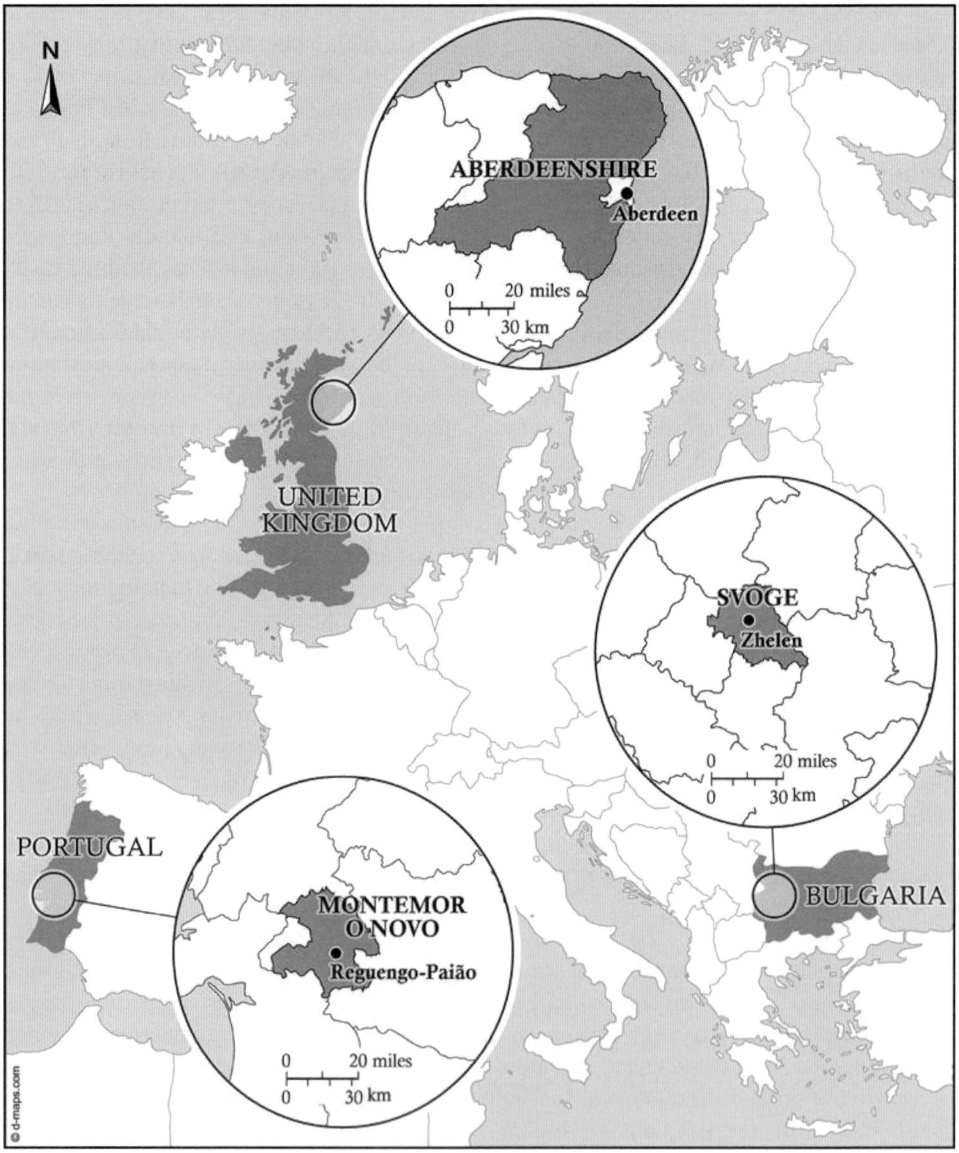

Fig. 5.1. Location of the three case studies: Reguengo-Paião, Zhelen and Aberdeenshire (Source: authors).

With regard to the actors involved in the niche, always land owners, three main categories were identified by the researchers: those who do not undertake any farming activity on their land; those who produce mostly for self-consumption or undertake recreational farming; and those who undertake commercial farming on a small scale. The diversity in the profiles of lifestyle farmers is also connected to their origins, which often influence their attitudes, and also undoubtedly to the networks that they relate to, and the way they interact with the other actors involved. Within the multi-level and multi-dimensional network of actors that characterize rural areas today, this networking profile and capacity play a crucial role (Ventura *et al.*, 2010). The complexity of this group of actors is illustrated by the lifestyle farmers found in the three case studies.

In the Scottish case, lifestyle farmers are predominantly new entrants to agricultural land management (such as people who have an urban job, in Aberdeen or elsewhere, and bought the farm as a place to settle). They may have an urban background or have an origin in the countryside but they have not usually inherited their land and, therefore, are new entrants to land ownership and farming. Although the more visible of these land managers (those that appear at agricultural events) are motivated towards small-scale livestock production and self-provisioning, there is a cohort who primarily purchased their properties in order to enjoy residence in an aesthetic environment, and a further group who purchased land in order to pursue recreational horse riding.

In the Portuguese case, the origins of lifestyle farmers are multiple and include both new entrants and individuals from local families. There are also urbanites who purchased land in the countryside in their quest for a rural lifestyle and are totally new to the area. Some new entrants, with urban or rural backgrounds, may be foreign, attracted by the southern European climate and lifestyle. There are also local people who had spent a considerable portion of their lives in an urban area but farmland has been kept in the family, and as they feel an attachment to their family place or an aspiration for better life quality, they settle in or near the family farm. These are considered here as returnees. Finally, there are locals who have always lived on a farm, which often used to be a main source of food or income for the family but the new generation have other jobs, nevertheless deciding to stay in the farm, as the farm lifestyle has improved and has become valued by society. A range of different combinations in between these major types also exist, generating high diversity and richness in lifestyle farming.

In the Bulgarian case, lifestyle farmers are represented by new entrants, small-scale farmers and volunteers. The most relevant are the new entrants (mainly young people/couples/families) who were not born in the area but have settled in the village and currently live there permanently. Their reason for settling in Zhelen is not agricultural production as an economic activity, but the connection with nature and expectations of a higher quality of life. These individuals typically inherited or purchased property and built new houses. There are also small-scale farmers whose origins lie in the region. Some of these individuals live permanently in the village, whereas others live there seasonally and cultivate their small farms at the weekend. All of these people produce agricultural products for their own consumption, occasionally selling their products to visitors and tourists. There are also volunteers who have an interest in rural lifestyles, want to develop closer contact with local people, or actively engage in agriculture and environmental protection on a temporary basis. They do not originate from the village and do not have their own land but take part in agricultural production and sometimes spend extended periods living on farms and sharing local lifestyles.

In contrast to the definition of a niche as being actively supported by its actors, it is evident from the study sites that lifestyle farming is not necessarily a formally organized process. As a result, there are no well-defined boundaries of the niche or actor groups. Although there are localized cases where formal initiatives are organized to pursue alternative ways of life (for example 'back to the land' movements such as the Bulgarian case studied in this research), in many cases it is simply individuals deciding that they would like to pursue lifestyle interests, including but not limited to food production, through a country residence (as in the British and Portuguese cases studied in this research). This shift in cultural orientation towards land use could be considered a landscape feature, as it exists outside of the commercial agricultural sector and, thus, the agricultural regime. Alternatively, it could be considered a feature of the leisure and housing regimes, interacting with the agricultural regime.

A further key tenet of the MLP is that niches require protection while they are developed. In all three cases, the niche has developed without protection, which is perhaps characteristic of innovations which are not technological in nature. Markard and Truffer's (2008) work on the MLP suggested two forms of niche: technological niches and market niches. Technological niches are most commonly described in MLP analyses, representing new technologies which have been created by actors and thus require protection until they become competitive. Market niches reflect 'natural anomalies' in regimes such as old technologies which flourish under new consumer preferences or application contexts (so that protection is not required). Lifestyle farming can therefore be considered a market niche; the value of agricultural land for recreational purposes and the entrance of new consumers have enabled the niche to flourish in the three sites.

Perhaps as a result of the lack of formal organization, the autonomy and diversity of actors involved is striking (Fig. 5.2). This diversity has made this process difficult to identify and define. What the actors of the niche have in common is the fact that their main motivation for keeping and managing their land is lifestyle, but beyond this, they may have little in common, even in the same local area because of their differing social status. While some lifestyle farmers in the Bulgarian and Portuguese cases may be representatives of the 'voluntary simplicity' way of living, this is not the case with the majority of lifestyle farmers in the Scottish case or with some Portuguese actors who retain their high urban incomes along with expensive lifestyle farming activities (e.g. buying land and houses, keeping horses).

The evidence seems to indicate that the more close connections there are in society with rural life, the more diverse is the profile of lifestyle farmers, as relations established with the countryside by individuals are multi-layered. In southern Europe, and particularly in the Iberia peninsula, society was largely rural until a few decades ago (Pais de Brito, 1996). The majority of the population lived on family farms, often subsistence farming, until the late 1960s, and the percentage of active population in agriculture was considerably higher than in western Europe until the 1980s. The population today largely retains its roots in the countryside; the close connection with the rural is just one or two generations behind. This is also mostly the case in Eastern Europe, even though recent history there is quite different. For example, the Bulgarian case shows that during the 1950s, production was organized in large state farms and family farms disintegrated, while industrialization, together with urbanization, developed. However, at the same time small-scale subsistence oriented farms have been maintained, and since the 1970s the number of family farms on private land has increased and become more widespread throughout the country resulting in the indirect urbanization of rural communities. This is definitively not the case in northern and western European areas, which have largely been urbanized for several generations. At the same time, southern European areas have, in the last two or three decades, undergone rapid modernization and become open to European markets, values and lifestyles. This explains why, in the Portuguese case, the move away from the countryside for an urban life, with associated extensification or land abandonment and as well as the transfer of land to older farmers is happening at the same time as newcomers are settling and new farming paradigms are put in place. There are complex family relations and multiple ways of accessing land, many of them grounded in the ownership of land by family members. The diversity of actors and motivations in the local area is high, as is the way that they interact with each other and their land. In the Scottish case, an example of a much more urbanized society, the types of farmers are more clearly differentiated and the separation of lifestyle farmers from the group of professional farmers is more straightforward.

Regime

In the MLP, socio-technical regimes are defined as the locus of established practices and rules (mainstream activities and their supporting institutions). According to Geels and Schot (2007), the regime accommodates a broader community of social groups and their alignment of activities, stabilizing existing trajectories in a diverse number of ways. Regimes are oriented towards fulfilling specific social functions. In this research we conceptualize conventional agricultural production as the primary purpose of the mainstream agricultural regime, recognizing that the agricultural sector has multiple functions. These functions include environmental conservation, which for these purposes is assessed as a separate regime. As shown in Fig. 5.2, the transition to lifestyle farming can be viewed as a case of multiple regime interaction: the real-estate regime in particular, in its relation with housing, is interacting with the agricultural regime in novel ways. Lifestyle farming can be located in the overlap between agriculture and real-estate regimes, as it fulfils the traditional function of residence provision for agricultural land managers, but the new 'lifestyle farmers' produce differently, and for different reasons. Lifestyle farming can also be related to the leisure regime, as it represents consumption of the experience of rural lifestyle (including, in some cases, interaction with livestock and horses); alternatively, the leisure orientation could be considered a feature of the changing socio-technical landscape. The move towards lifestyle farming may reflect the landscape trend of increasing disposable income and societal shifts towards the consumption of 'experience'. It is also related to the landscape trend of urbanization, not so much as a physical manifestation of growing cities but as an increased influence of urban values and lifestyles in the countryside.

The three key regimes with which the lifestyle farming niche interacts are therefore: agriculture, real-estate and conservation (Fig. 5.2). The first two regimes have always interacted, as farms (at least small-scale family farms) have also been living places; the real-estate regime also embraces the network of social relations established in a local community that share many everyday life practices and events. Currently, and as a response to modernization of agriculture, there is involvement with a third regime: the conservation regime, which aims to maintain the landscape and environmental assets of these areas. The small-scale farms considered in all three cases have been until relatively recently, both living places and production units, being production-oriented for the market or for self-consumption. Several of the farms which have not recently changed owners are still managed within this paradigm, while at the same time in Portugal and Bulgaria, residents are, or have been until recently, part of a community, a village with intense social relations amongst members.

Several sub-regimes constitute the three main regimes: (i) food production, processing and marketing; (ii) rural housing (farms as places to live) and the real estate market; and (iii) consumption of countryside amenities and rural lifestyle. Some of these sub-regimes overlap, and not all are equally important in each case. Moreover, other sub-regimes are important in the different cases. Examples include cultural heritage from agriculture in the case of Bulgaria; traditional exploration of wild resources in the case of Portugal; or the horse industry in the case of Scotland.

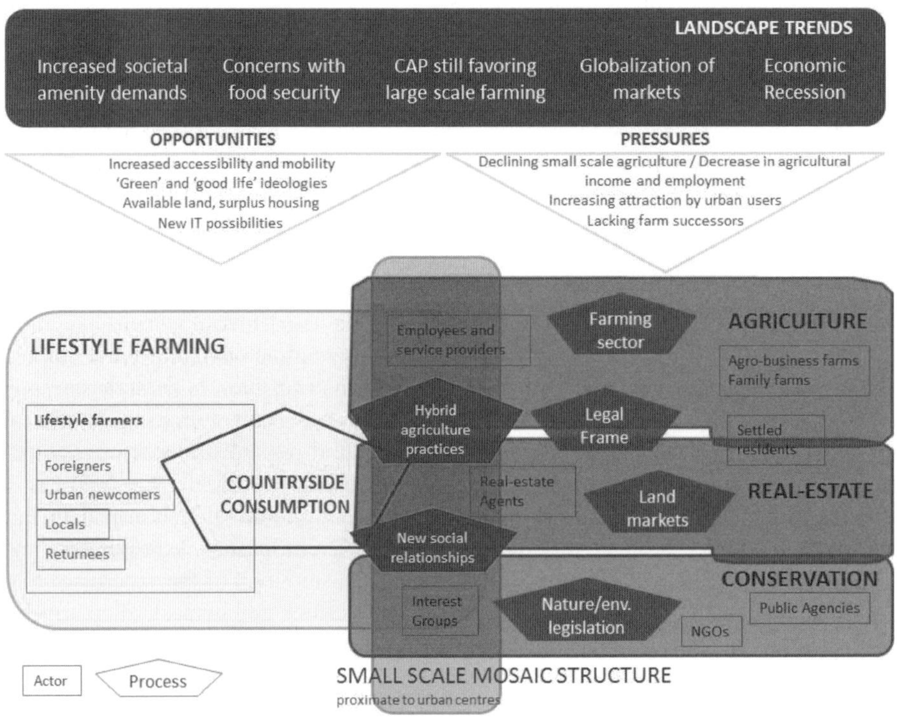

Fig. 5.2. Key actors and processes at the niche, regime and landscape levels, and their interactions in transitions to lifestyle farming (Source: authors).

The dominant regime with which the niche interacts is agriculture. Modernization in agriculture is a critical trend in the dominant regime. In Scotland, the process of modernization started earlier and this means that a number of farmers have shifted from full-time to part-time farming. This change is generally due to high capital investment and agricultural specialization, as well as to the increasing size of plots, machinery and buildings and the need to compete in globalized markets. Individuals who grew up on farms often cannot afford to start their own farms and may work elsewhere, adopting small-scale lifestyle farming as a way of maintaining their heritage (but without the need to make a living from it). Commercial farming tends to focus on large farm units, while there is an increase in the demand for small farms for uses other than production. In all three countries, the impact of agricultural modernization is evident in a decrease in the number of farm units, abandonment of production on land less suitable for agriculture, or a shift from market-oriented production to self-sufficiency. Also, in Bulgaria and Portugal small-scale agriculture has been downsizing markedly, accompanied by an increase in the average size of farm units, and the small-scale agriculture remaining is mostly carried out by a local, ageing population. Very low agricultural wages and low social value of farming also characterize recent trends in all three cases.

In recent years, investment in land in the study areas has been increasing. This is due to the amenity value of the land, for both permanent and for secondary residence (week-end and holiday). This is related to the small size of farms, their difficulty engaging in the globalized economy, the landscape quality of the area, and the relative proximity to a large city (Sofia, Aberdeen and Lisbon). In this way, rural housing and real estate markets have become important sub-regimes.

Two other trends have strong impacts and reinforce these changes to small-scale farms in all three cases (i) increased accessibility to rural areas from urban centres and increased mobility of individuals; and (ii) technological innovation in computers, internet connections and mobile phones, which has created a new structure of connections and opportunities for long-distance work.

Further, an important aspect shaping land use in lifestyle farming areas, particularly in the Portuguese and Bulgarian cases, is often related to nature conservation and environmental legislation, and to the interests of associated actors. New houses cannot be built next to existing farm buildings due to planning restrictions, which increases pressure on farms and farm buildings. While on one hand this may appear to constrain the development of the niche, it is in fact maintaining important attractiveness factors in the area (landscape,environmental quality, and low population density).

Finally, to some extent, lifestyle farming can also be related to the leisure regime, as it represents consumption of the experience of rural lifestyle (including, in some cases, raising livestock and horses).

Landscape

The analysis of these cases demonstrates the importance of landscape factors in creating an enabling environment for change. Within the MLP conceptualization, landscape is defined as socio-technical: the long term exogenous trends at the macro level (Darnhofer, this volume). These trends include demographics, political ideologies, social values, macro-economic patterns and climate change. Within the MLP, landscape level factors may destabilize the regime, exposing systemic problems or contradictions and creating 'windows of opportunity' for niches to influence the regime and for the mainstreaming of niche innovations. In the case of lifestyle farming, while these landscape factors (Fig. 5.2) clearly had an influence – the change in societal values towards an 'experience' culture, for example – the territorial and spatial dimensions played a key role in the locations where lifestyle farming has occurred. It also appears that the agricultural regime itself was gradually destabilized by low commodity prices and farming incomes, which led to low land prices and pluriactivity, opening the door to recreational land use by (comparatively) wealthy ex-urbanites in these regions.

Fig. 5.3 shows the eight main factors which have been identified by the authors as contributing highly, in different combinations, to the attractiveness of a rural area for lifestyle farming. These attractiveness factors express the overlap of landscape trends (and the opportunities and pressures it creates) and regime traits, and therefore are an expression of the complex interrelations between the two levels. Several of these factors address the physical characteristics of the land itself. The small-scale mosaic structure of farms reduces their utility for commercial farming, and creates conditions for the farm to be managed as a secondary activity. This was present in all three cases. Both the Bulgarian and Portuguese cases were also located in regions of low potential for intensive and competitive agricultural production, due to biophysical constraints, thus further limiting prospects for commercial farming. Although the Scottish case study was set in one of Scotland's best agricultural regions, owing to the northern latitude, the area does not have the highest quality agricultural land. Lower quality agricultural land is often located in areas with high 'amenity appeal' (e.g. near mountains), which also facilitated the development of the three cases. Strong local farming knowledge cultures were particularly important in the Bulgarian and Portuguese cases, where food production was more emphasized. Active cultural life also holds appeal for newcomers and returning residents. All three of the cases were within commuting distance of large urban centres, and in the Scottish and Portuguese cases, no legal framework required the use of agricultural land for agricultural production, thus leaving the land available for personal recreational use.

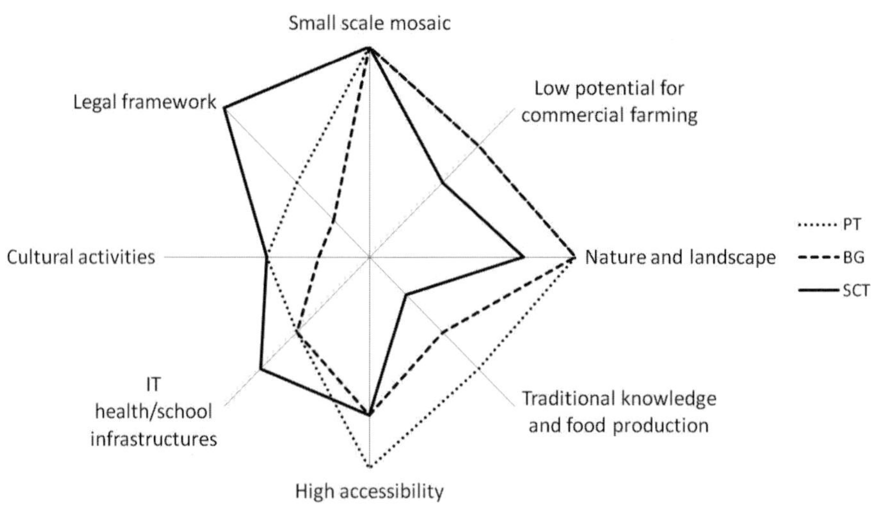

Fig. 5.3. Comparative analysis of the factors that play a role in the attractiveness of a rural area for lifestyle farming (Source: authors).

The difference between lifestyle farming and other niches typically studied with the MLP, is that lifestyle farming represents a socio-technical transition where the social aspects are predominant. Lifestyle farming has a technological component; recent advances in IT and infrastructure have made it easier to pursue lifestyle farming, while retaining employment and urban social networks. In all three cases, these supporting technologies were important to the development of lifestyle farming but it is the new ways of thinking about the countryside, and associated beliefs and values, which are driving the transition.

Is lifestyle farming anchoring?

Lifestyle farming is becoming a mainstream activity in the regions studied, in the sense that those living on farms or providing services within the agricultural sector will be aware of it. Lifestyle farming impacts economically, in terms of the capital available and used within the countryside, rather than productively (in terms of agricultural commodities produced). It also has a social impact on rural communities. The practice of lifestyle farming is sufficiently diverse that it is difficult to detect specific changes which reflect mainstreaming – these would more likely reflect societal trends generally (for specific types of lifestyle activities, for example keeping a few chickens, wearing specific clothing, horse riding) than trends that would push the lifestyle farmers closer to conventional farming.

As explained in Darnhofer (this volume), 'anchoring' refers to emerging forms of linking between the niche and the regime, while 'linking' expresses the more robust interaction at later stages. Anchoring within the regime has been driven by two trends: (i) the productivist quest for economies of scale and competitiveness in the global market, which left the small-scale farms outside what is considered competitive farming and therefore as 'surplus' farmland; and (ii) the cultural shift towards valuing leisure lifestyles and 'green' amenity. In particular, it reflects the commodification of nature and the rural – land is valued more highly for recreational and housing use than for agricultural production (Phillips, 2005). The anchoring process has been achieved largely through the purchasing power of the new lifestyle farmer managers, rather than through active lobbying by proponents. As a niche, it did not require protection from markets; instead it could be viewed as a product of real-estate regime access to unprotected agricultural land markets.

Lifestyle farming has anchored into the agricultural regime in several ways. Lifestyle farmers purchase land from the same vendors, and sometimes inputs from the same suppliers, as commercial farmers. This is evidence of economic institutional anchoring: both types in recent years have re-oriented their activities to be able to support this growing market. In Scotland, estate agents have divided up farms for sale into residential and commercially-scaled properties, thus enabling the creation of more lifestyle-scaled units. Input suppliers have developed products specifically for the (wealthy) lifestyle markets: smaller scale machinery and quantities of seed and feed, even introducing new clothing ranges. This is not so evident in the Portuguese and Bulgarian cases but lifestyle farmers there are also purchasing their products and equipment in the local stores, adding to the stores' business and supporting the diversity on offer. Lifestyle farmers do not typically have large amounts of equipment but may considerably boost purchases of small-scale equipment like lawnmowers and hedge trimmers. They are also more likely to make more use of farming services (for example contractors, veterinarians) than commercial farmers, thus providing a financial basis for on-going service provision.

There is some evidence of 'network anchoring'. Elzen *et al.* (2012) define network anchoring as the changes that occur in the network of actors that produce, use and develop a novelty. Lifestyle farming typically involves the engagement of new actors into the agricultural sector, arguably forming 'new' networks. There is also some evidence that these new actors are penetrating existing networks within the agricultural regime. For example, in the Scottish case study, lifestyle farmers often participate in specific networks associated with commercial farming, such as rare breed societies or pedigree livestock shows. In the Portuguese case, they employ or buy services from conventional farmers or from other local conventional service providers and moreover participate in local NGOs such as cultural collectives or hunting groups. However, they remain largely disconnected from mainstream agricultural knowledge systems (such as advisory services), and although they may utilize commercial marketing venues, they are not typically part of the associated networks that have built up at these locations. In contrast, the new lifestyle farmers in the Portuguese site appear to integrate with more traditional networks, whereas in the Bulgarian site, participants are part of a 'movement', which represents a network in itself, albeit separate from local networks.

What is perhaps most notable about the anchoring of countryside consumption is the lack of normative institutional anchoring; formal or informal rules about what is desirable which can be embedded in laws, regulations or policies (Elzen *et al.*, 2012). Lifestyle farming is largely unrecognized in agricultural policy in all three study sites. Instead, policies are clearly oriented towards commercial production. As a result there are several unintended influences on the evolution of lifestyle farming. Tax advantages associated with managing agricultural land, intended to assist commercial farmers and their successors, can also be of advantage to lifestyle farmers. In Portugal and Scotland land is for sale to the highest bidder, making it easily transferable from commercial to lifestyle use. However, legal reporting requirements (for instance livestock tracking and welfare reporting) are also designed for commercial-scale farming operations and can act as a barrier to less intensive, leisure-oriented management of livestock. Lifestyle farmers are often excluded from traditional sources of state support (such as agri-environmental funding) through lack of awareness. Despite mainstreaming, lifestyle farming thus continues 'under the radar' of official state practices.

Technological anchoring is also not immediately obvious in the analysis of lifestyle farming. Some new specific technological approaches are developed by, or for, lifestyle farmers, for example in order to facilitate the leisure management of livestock or to obtain high quality or quantity production in small plots (such as for fruit and vegetables). These technological novelties may also interest large-scale commercial farming, creating a link. However, this niche is not primarily based in technology but as an ideal of life quality and for the use of physical space, although further definition and refinement of this ideal is difficult to assess. The wide range of proponents of the ideal – and its widely varying forms – suggest that it is mainstreaming through becoming more flexible and amenable to social standards, rather than through becoming more specific.

Discussion

The increasing consumption of the countryside is an ongoing process, operating at different spatial and temporal scales throughout Europe. Analysis utilizing the MLP was found to be useful for identifying who is involved in change, and why the processes occurring are not always explicit. The particularities of the observed niche also bring new insights into the MLP conceptual construction. The analysis reveals resistance from the regime as well as contradictions within the niche. It reveals also how the niche can change over time according to the interplay of its own actors but also the changes in the external context; the landscape. Starting as a consumption-driven land management style, lifestyle farming is in some cases moving toward a stronger role in production, responding both to actors' aspirations and to the global crisis affecting food markets and family incomes and may yet be redefined again. This suggests, for the MLP, that the establishment of the niche phase more often than not may overlap with the take-off phase of the transition, the latter contributing to defining the contours of the niche.

Our analysis is consistent with a criticism raised by Lawhon and Murphy (2012) of the lack of attention to space and scale within the MLP. Local physical landscape features as well as the location in relation to large urban areas clearly have an impact on where lifestyle farming is most likely to occur. Characteristics of the locale thus determine how, and if, this niche will occur. In addition, the changing characteristics of this space, as a result of increased transport and IT facilities, have been shown to be crucial. This issue is of particular importance to studies of agricultural transitions as these, more than most transitions, will be rooted in biophysical space (see also Darnhofer *et al.*, this volume).

In this way, the role of land in transitions in agriculture needs to be questioned. Differing from other transition processes, in agriculture the issue of land is part of the process. It could be that, within the MLP, land itself is to be considered in a similar way as technology whereby the use is developed and shaped by different actors for different purposes. The emergence of the niche also has particular contours here. The landscape pressures and opportunities enabling niche emergence have been described and are represented in Fig. 5.2. They correspond to long term changes in society and not to a window of opportunity created by a sudden event, although the food crisis of 2007–2008 and the recent global financial crisis may have fuelled them further. As such, these landscape pressures and opportunities have not resulted in awareness in the dominant regime (agriculture). So there are a lack of strategies and policies that acknowledge the phenomena of lifestyle farming, while at the same time the regime also does not oppose it. It may also be that this niche will, in any case, not be best supported by a transition of the regime, but rather by creating its own parallel space which acknowledges the diversity of land management paradigms emerging all over Europe and deviating from production as the sole priority.

The agricultural regime has, in the last few decades, been concerned with a totally different agenda, namely the increase in industrial production and globalized networks, as well as problems related to market fluctuations. Therefore, it has remained weak in relation to processes affecting small-scale farming. The real-estate regime is becoming dominant, probably also due to this weakness, penetrating the agricultural regime in terms of decisions over land. With the higher economic power of those able to buy land for lifestyle farming, and the dynamics of real-estate agents, there may be a power shift occurring, where in particular places, the real-estate regime replaces the agricultural regime in shaping the context for the management of agricultural land.

The assessment thus demonstrates that the agricultural regime is missing two major opportunities. The first would be to strengthen its territorial role and responsibility for the management of natural resources as well as the physical space in rural areas – a physical planning role for the rural, which is strongly needed in the light of increasing resource scarcity and land use conflicts. This role has been present in the political discourse of the agricultural regime in Europe but to date it has been weak in content – as demonstrated by the increased dominance of the real-estate regime in the cases studied. The second would be to embrace a leading and overarching role in supporting innovation by embracing the local autonomy and food quality agendas, reinforcing European diversity and regional specificities in the face of global production chains. This would support differentiation in relation to world agriculture and, therefore, could contribute to competitiveness.

Concerning actors, the analysis also has shown how, in relation to the regime, a non-organized niche acts differently to a niche resulting from collective action and a shared intention to change the established regime. The three cases have shown that the changing capacity of these actors is based on their economic power and their human capital, combining higher education, entrepreneurship and often strong convictions. It may be the case that an incremental change within the housing industry and real estate is becoming a radical transition within agriculture.

Conclusion

Despite the acknowledgement of increased countryside consumption in the literature on rural space and rural society, lifestyle farming remains a relatively unseen phenomenon by those involved at the regime level. As has been discussed above, the agricultural regime is not concerned with this niche. Those managing the real-estate sector are taking decisions over land, and acknowledge the value of land due to its amenity interest but they do not consider the related production activity. Only the conservation regime, which is by far the weakest presence here, recognizes the importance of maintaining production and securing sustainability. The conservation regime promotes the environmental agenda, and the production of goods and services, which will in turn favour the real-estate regime but the conservation regime does not have the power to significantly impact on the other regimes.

Finally, lifestyle farming remains primarily an unseen process. These are unseen farmers and they seem to remain as such; they *are* seen as community members, property owners and care-takers of the physical landscape, but not as farmers and food producers, and not as active actors in promoting new production paradigms. Despite their potential to enhance regional sustainability in many local areas in Europe, lifestyle farmers therefore remain unsupported by policy actors.

References

Elzen, B., van Mierlo, B. and Leeuwis, C. (2012) Anchoring of innovations: Assessing Dutch efforts to harvest energy from glasshouses. *Environmental Innovations and Societal Transitions* 5, 1–18.

Darnhofer, I. (2015) Socio-technical transitions in farming. Key concepts. In: Sutherland, L-A., Darnhofer, I., Wilson, G.A. and Zagata, L. (eds) *Transition Pathways towards Sustainability in Agriculture: Case Studies from Europe*. CABI, Wallingford, UK, pp. 17–32.

Darnhofer, I., Sutherland, L-A. and Pinto-Correia, T. (2015) Conceptual insights derived from case studies on 'emerging transitions' in farming. In: Sutherland, L-A., Darnhofer, I., Wilson, G.A. and Zagata, L. (eds) *Transition Pathways towards Sustainability in Agriculture: Case Studies from Europe*. CABI, Wallingford, UK, pp. 189–204.

Geels, F.W. and Schot, J. (2007) Typology of socio-technical pathways. *Research Policy* 36, 399–417.

Holmes, J. (2006) Impulses towards a multifunctional transition in rural Australia: Gaps in the research agenda. *Journal of Rural Studies* 22, 142–160.

Holmes, J. (2012) Cape York Peninsula, Australia: A frontier region undergoing a multifunctional transition with indigenous engagement. *Journal of Rural Studies* 28, 1-14.

Lawhon, M. and Murphy, J.T. (2012) Socio-technical regimes and sustainability transitions: Insights from political ecology. *Progress in Human Geography* 36, 354-378.

Markard, J. and Truffer, B. (2008) Technological innovation systems and the multi-level perspective: Towards an integrated framework. *Research Policy* 37, 596-615.

Marsden, T. (2013) From post-productionism to reflexive governance: Contested transitions in securing more sustainable food futures. *Journal of Rural Studies* 29, 123-134.

Milone, P. and Ventura, F. (eds) (2010) *Networking the Rural. The Future of Green Regions of Europe*. Van Gorcum, Assen, the Netherlands.

Murdoch, M., Lowe, P., Ward, N. and Marsden, T. (2003) The differentiated countryside. In: Lowe, P., Marsden, T., Murdoch, J. and Ward, N. (eds) *The Differentiated Countryside*. Routledge, London, UK.

Pais de Brito, J. (ed.) (1996) *O Vôo do Arado*. Museu Nacional de Etnologia, Ministério da Cultura, Lisboa, Portugal.

Phillips, M. (2005) Differential productions of rural gentrification: Illustrations from North and South Norfolk. *Geoforum* 36, 477–494.

Renting, H., Rossing, W.A., Groot, J-C., van der Ploeg, J.D., Laurent, C., Perraud, D., Stobbelaar, D.J. and van Ittersum, M.K. (2009) Exploring multifunctional agriculture. A review of conceptual approaches and prospects for an integrative transitional framework. *Journal of Environmental Management* 90, S112–S123.

van Berkel, D.B. and Verburg, P.H. (2011) Sensitising rural policy: Assessing spatial variation in rural development options for Europe. *Land Use Policy* 28, 447–459.

van der Ploeg, J.D. (2009) Transition: Contradictory but interacting processes of change in Dutch agriculture. In: Poppe, K., Termeer, C. and Slingerland, M. (eds) *Transitions Towards Sustainable Agriculture and Food Chains in Peri-Urban Areas*. Wageningen Academic Publishers, Wageningen, the Netherlands, pp. 293–307.

van der Ploeg, J.D. and Marsden, T. (2008) *Unfolding Webs. The Dynamics of Regional Rural Development*. Van Gorcum, Assen, the Netherlands.

Ventura, F., Milone, P. and van der Ploeg, J.D. (2010) Understanding rural development dynamics. In: Milone, P. and Ventura F. (eds) *Networking the Rural. The Future of Green Regions of Europe*. Van Gorcum, Assen, Netherlands, pp. 1–29.

Chapter 6

More than just a factor in transition processes? The role of collaboration in agriculture

S.R. Schiller[1], C. Gonzalez[2], S. Flanigan[3]

[1]*Institute for Rural Development Research, Frankfurt am Main (schiller@ifls.de);*
[2]*University of Évora;* [3]*James Hutton Institute, Aberdeen*

Introduction

The multi-level perspective (MLP), which has been applied in the FarmPath project (Darnhofer, this volume), implies that collaboration between actors is essential for niche development; and likewise collaboration between niche and regime actors needs to take place in a transition process. Studies on transition have, in general, noted the importance of these interactions for the transition process. However, there is little knowledge on the development of collaboration and its role in the context of transition. Using three initiatives as case studies, this chapter will provide insights into collaboration in agriculture, and its role in the transition towards regional sustainability of agriculture.

Collaboration, as opposed to conflict and competition, is a major pattern of human behaviour necessary for the production of goods and services (Fuchs-Heinritz and Barlösius, 2010). Individuals act together within a common context to achieve a common objective. From this general definition, it becomes clear that collaboration can take very different forms in terms of the number of actors involved; the intensity of interaction; the activities undertaken; and the social norms and values, framework conditions and objectives pursued. Furthermore, collaboration can be formal and/or informal. Cooperatives, particularly agricultural cooperatives, are a distinct organizational and regulated form of collaboration, generally with a solidarity-oriented, rather than profit-oriented approach. In agriculture, two key forms of collaboration can be differentiated: 'vertical cooperation' between agricultural producers and other businesses in sectors (for example, suppliers and processors); and 'horizontal cooperation' between agricultural producers (such as machinery rings) (Klischat *et al.*, 2001).

Collaboration has also been identified as important in relation to ensuring the social and economic sustainability of farming. Sutherland *et al.*a (this volume) denoted multiple functions of agriculture which are identified in both literature and EU agricultural policy as key to sustainability in agricultural systems. Agriculture is expected to provide a range of goods, functions and ecosystem services from food security to employment; rural development; soil and water quality; nature conservation; and amenity functions. These expectations make it increasingly difficult for farmers and the agricultural production chain to adapt to the changing needs and preferences of consumers, citizens and policies. It

 Transition pathways towards sustainability in agriculture: Case studies from Europe. (eds L-A. Sutherland *et al.*)

furthermore shows that it is unreasonable for individual farms or farming systems to attempt to meet all of the above demands alone. Instead, it is proposed that sustainability is best addressed by enabling actors within a region to optimize their own specific opportunity sets (for example from natural resources, farming systems, social capital, governance structures, infrastructure and economic development) in coordination with other relevant regional actors.

Regional sustainability has to be grounded in the local context and implies regional differences in the normative goals of the transition towards sustainability. As a consequence, there is a need for interactions between individual farm models and farming systems at the regional level, as a key aspect of sustainability: what will be decisive is whether the farm models present within a region can interact to meet these changing needs.

From a transition management perspective, collaboration can be identified as a 'socio-technical' innovation. This innovation forms the basis of the three niches studied here, illustrating how novel forms of collaboration attempt to break through the regime and promote transitions. Referring to the MLP, collaboration may be relevant in, for example, supporting the development and establishment of niches (in the pre-development stage of sustainability transitions), in potentially anchoring the niche in the regime, or in accelerating the transition.

This chapter will show how collaboration can be conceptualized as a socio-technical innovation, based on the in-depth analysis of three initiatives (machinery rings in eastern Scotland, CRIE Montado in Portugal and Regionalwert AG in Germany). Furthermore, it will explore the role of collaboration in transition processes towards the sustainability of agriculture, and summarize key factors influencing collaboration and transition.

Collaboration in transitions towards sustainability of agriculture

Collaboration between different agricultural actors has been highlighted as important for the transition process in many of the initiatives studied in the FarmPath project (see, for example, Peneva *et al.*, this volume). The three initiatives discussed in this chapter were chosen on the basis that the collaborative aspect was among the key elements making them innovative. Each of the initiatives has a different organizational form and integrates actors from a variety of sectors:

- In Portugal, the 'CRIE Montado' initiative is a small group of agricultural entrepreneurs collaborating for multifunctionality in the Montado, and promoting the development of multifunctional farms and agriculture through experience and information sharing, thereby extending farm activities from agricultural production to tourism and recreation, education, social inclusion and product transformation.
- Formalized resource sharing has been studied in an investigation with two Scottish machinery rings (Ringlink and Borders Machinery Ring). Machinery rings provide opportunities for increased efficiency in agriculture by offering a low-cost mechanism for resources, such as machinery and labour, to be shared according to supply and demand, which can help farmers to reduce costs and provide an alternative means to generate additional income. Commodities trading and training provision have also become part of the range of services that machinery rings offer to members.

- In Germany, the 'Regionalwert AG' operating in the Freiburg region is a citizen's shareholder corporation that facilitates access to shareholders' capital for organic farms and businesses. As such, it follows the objective of creating regional value not only in financial terms but also in social and environmental terms.

Table 6.1 provides an overview of the initiatives' regional affiliations, different organizational forms, purposes, and the number and types of actors involved.

Table 6.1. Characteristics of the three collaboration initiatives (Source: authors).

Initiative	CRIE Montado (PT)	Machinery Rings (UK)	Regionalwert AG (DE)
Region and approx. size	Montemor-o-Novo, 1,200 km^2; Alcácer do Sal, 1,500 km^2	North East Scotland, 9,000 km^2; Scottish Borders, 4,500 km^2	Administrative district of Freiburg, 9,000 km^2
Organi-zational form	Informal group of agricultural entrepreneurs	Legal agricultural cooperative	Citizen shareholder corporation
(Formal) initiation	2009	1987	2006
Activities	Information and experience sharing, lobbying	Sharing of resources, information sharing, commodity exchange	Investment into organic businesses through shareholders, information sharing, commodity exchange
Actors	Farmers, processing, marketing	Farmers, supplier businesses (commodities, services, etc.), organizations, households	Shareholders, partner businesses: farmers, production, processing, marketing, wholesale, retail, catering
No. of actors (in 2012)	10 to 15	2,600 (Ringlink); 860 (Borders Machinery Ring)	16 partner businesses, 460 shareholders

In all three initiatives, collaboration is a response to different pressures. Whilst economic pressures were common to all three initiatives, the difficulties in acquiring capital for investment in farming added to issues of farm succession (and economic viability) in the case of Regionalwert AG in Germany. Other pressures included specialization of production chains, sectoralization of administration and markets, and competitiveness of global markets.

As discussed above, collaboration is regarded as one of the basic principles of human behaviour, thus, collaborative action as such is not innovative. From a transition perspective, however, these three initiatives represent new forms of collaboration and can be identified as 'socio-technical' innovations as they have introduced new practices, network connections or technology in their particular collaboration forms. CRIE Montado (Fig. 6.1) innovated by creating a regional thematic group dedicated to implementing multifunctionality and 'green' rationales in the Montado system through a multi-level (national, EU) model of collaborative learning, networking and lobbying.

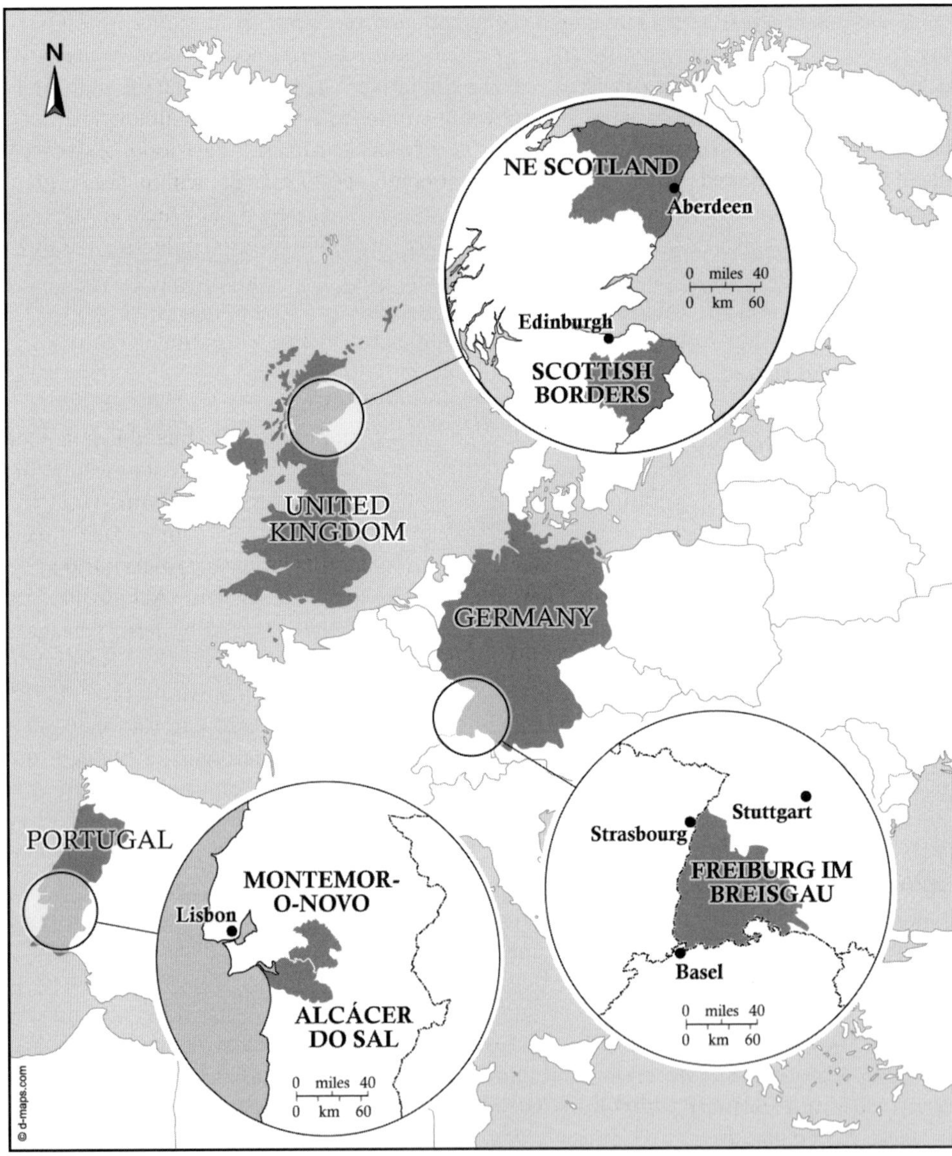

Fig. 6.1. Location of the three case studies: Montemor-o-Novo, eastern Scotland and Freiburg im Breisgau (Source: authors).

The Montado is an agri-silvo-pastoral system predominant in the western Iberian Peninsula (*dehesa* in Spain). It is a 'triple' complementary land use, adapted to low soil fertility and to the Mediterranean climate, and consists of open, evergreen forest, livestock grazing and cultivation. This management system results in a savannah-like landscape (Pinto-Correia, 1993; Surová and Pinto-Correia, 2008).

In the Scottish context, machinery rings have been introduced as an innovative mechanism in terms of institutional arrangements, but they have also introduced technical innovations to facilitate sharing of resources amongst farmers. What makes the Regionalwert AG socio-technically innovative in terms of collaboration is the newly created (financial) linkage between shareholders and partner businesses: although distant, the initiative enables shareholders to invest in organic farms and agri-food businesses, meeting ecological and social requirements (organic certification) within the region. Another innovative aspect of this particular initiative is its valuation which is not only based on financial returns but also on social and environmental returns, reflecting the three pillars of sustainability.

CRIE Montado: Collaborating for multifunctionality in the Montado silvo-pastoral system (Portugal)

In October 2008 an international network was formed at the European Eemland Conference 'The European Versatile Countryside'. The network is dedicated to disseminating the collaborative model in Jan Huijgen's conceptualization of multifunctional farms (Postema and Ziel, 2008), put in practice in his Centre Eemlandhoeve. The collaborative model firstly links local/regional thematic groups (e.g. milk producers, Montado owners) and then networks them with existing groups at national and European levels. Acknowledging multifunctionality is extremely hard to implement and cannot be faced by farm managers alone, Huijgen's vision and associated collaborative model proposes to strengthen multifunctional farms through multi-level networking and in this way, provide: (i) reinforcement and mutual support; (ii) collaborative learning and sharing of information on legislation, administrative processes, funding and markets; and (iii) lobbying and communication as an interest group. Multifunctionality demands high versatility of farms and farmers, and a high level of detailed knowledge and know-how in multiple activities and specialized sectors at the same time, which can be addressed through collaborative approaches.

One of CRIE Montado's members was part of the process and united a group of 12 farm entrepreneurs from his personal and professional network to establish CRIE Montado in Portugal, based on geographical (Montemor-o-Novo and Alcácer do Sal) and interest affinities (multifunctionality, environmental sustainability, the Montado system). A collaboration agreement was made in February 2009, where the privilege of informality and flexibility was defined. Members of the initiative met at informal meetings, where they shared ideas and discussed difficulties and solutions. Some joint visits and training sessions were conducted, as well as meetings with a few key regional and national institutions. Through collaboration, information sharing and learning, experiences in diverse activities suitable for the Montado system (from organic farming and cork exploitation to leisure, tourism, rehabilitation centres, trekking and nature conservation activities) could be shared amongst members. As a result of this process, new economic relationships were created, such as joint marketing activities or trading between members. Besides this, the group promoted greater flexibility and sensibility towards territories with similar conditions on a few occasions.

However, as a result of several external and internal factors, including the financial crisis and problems with the way that the collaborative model worked, from mid-2010 onwards, group activities declined. In March 2012, a new meeting was held, re-starting the sequence of regular informal meetings and a number of new members joined whilst several older members left.

Formalized machinery and labour sharing: Machinery rings in Scotland (UK)

The Scottish machinery rings developed on the basis of successful implementation of the concept in other European countries, such as Germany and France. Facilitated by the Scottish Agricultural College (SAC), machinery rings evolved from small-scale local discussion groups and have become an established actor in the context of rural areas. The first ring was established in the Scottish Borders in 1987 as a means for farmers to reduce their fixed costs by sharing machinery and labour. Over the past 25 years, they have broadened their membership (to include trades, commercial companies, other organizations, and households) and evolved to include a range of additional functions, including training provision, commodities trading and support for renewable energy development. There are currently nine rings in existence across Scotland. Since they were established, membership numbers have grown (and fallen) at different rates in different regions based on a variety of factors, including increased awareness, introduction of new services and amalgamations/ expansions of area. Members interact with the machinery ring organization on a more or less frequent basis as services are required or desired, predominantly by telephone or email. Annual general meetings are held, providing members with an opportunity to interact with the organization and the board.

Key features that contributed to the progression (and success) of machinery rings as an emerging niche include specially developed computer software and early adoption of direct debit technology. Both technological advancements brought increased ease and efficiency to new transactions in the agriculture sector. Processes instituted by the rings have also resulted in greater capacity for supply-demand relationships in agriculture to be managed in a more strategic way at a regional level. Through creation of an extensive membership base, and in some instances formalizing existing relationships between farmers, increased formal collaboration has been an important societal process. Leadership by, and within the rings, is also a significant contributing factor in the rings' progress, and includes good management provided by staff, and strategic direction provided by boards of farmer directors. A further defining factor in the success of machinery rings relates to providing a mechanism for farmers to collaborate in a way that is ultimately voluntary (unlike other forms of agricultural cooperatives in Scotland), so they can also retain a desired level of independence and autonomy in their farm business.

While niche actors sought to introduce machinery rings to Scotland, there is no suggestion that their intention was to initiate a regime-level transition at the outset and the unpredictable nature of transition has entailed elements of success and failure during the 25 years since their introduction. Many of the seven 'lessons' Roep and Wiskerke (2004) associate with stabilization can be related to machinery rings in Scotland and can often be associated with their leaders' awareness and tenacity. For example, connections made by machinery ring managers and board members with actors and processes in the wider regime, are significant in terms of securing their position within it. Appreciation that machinery rings are involved in a continuous process in terms of aligning strategies and expectations (such as evolution into different markets and being responsive to changing policy), has helped to broaden their value. Finally, ensuring that machinery rings provide a mechanism that allows actors to improve their own situation has resulted in widespread participation across the sector. In this respect, collaboration in machinery rings provides an alternative which ultimately fits within the still widespread productivist ideologies that characterize agriculture in Scotland.

Fundamentally, by challenging the mostly individualistic way things are done within the agriculture regime, machinery rings have offered a viable alternative for a large number of processes and interactions. Formation of a large 'internal' machinery ring network has been an important aspect of this. In addition, alliance and integration with other regime level actors and networks (such as the National Farmers Union) and national level processes (for example policy consultations) have resulted in the presence of machinery rings becoming established within the wider agricultural regime.

Citizen shareholder capital for regional value creation: Regionalwert AG Freiburg (Germany)

The Regionalwert AG was formally founded in 2006 by a charismatic farmer who inherited and managed an organic (bio-dynamic) horticultural business in the municipality of Eichstetten. The main objective of the initiative was to provide access to capital for young farmers and new entrants within the region while at the same time increasing citizen participation in agri-food businesses. The establishment of the initiative was preceded by an extensive period of conceptual development with regional stakeholders, which could be traced back to discussions following the Chernobyl catastrophe in 1987. Part of that was a series of events called 'culture in the greenhouse' organized by the founder from 1998 onwards, which led to initiation of a foundation to support the breeding of local plant cultivars as part of the Agenda 21 process in 2001. The Regionalwert AG was officially founded with two partner businesses and around 1.7 million Euros of equity capital. The capital was increased to almost 2 million Euros in 2010. Only certified organic businesses are eligible for financial support through the Regionalwert AG. This support could take the form of: (i) purchase and lease of land or entire farms; (ii) silent participation (AG is a financial donor and accepts profits or losses but has no other influence); (iii) partnership shares; or (iv) shareholder loans. By 2012, the network of supported farms and businesses had increased to a total of 16 businesses comprising six farms, two processors, four retailers/wholesalers and four service providers. Around 460 shareholders had invested in the 16 partner businesses.

Moreover, the Regionalwert AG aims to increase collaboration between the partner businesses to strengthen regional value chains. Compared to the situation of very limited collaboration before the start of the initiative, collaboration between partner businesses has developed or intensified through the Regionalwert AG. The degree to which this has been achieved is variable. For example, a dairy farm and a neighbouring horticultural business are collaborating closely, sharing and exchanging resources and labour; food processors and catering services are also sourcing some of their raw materials from partner businesses. However, there is more potential to be realized, to increase collaboration between existing farms, retailers and processors, in terms of commodities as well as services and knowledge. Infrastructure for logistics has been mentioned as one of the core issues to be solved in order to intensify collaboration. There are usually no direct interactions between shareholders and the partner businesses; however, there have been ideas about ways to intensify collaboration beyond financial relations. To facilitate mutual learning, which is another core objective of the initiative, a platform for knowledge exchange between the businesses has been institutionalized in the form of a bi-monthly entrepreneur forum.

Besides the innovative financial collaboration between citizens, shareholders and partner businesses, another innovative objective which is an integral part of the Regionalwert AG concept, is regional sustainability. Regional sustainability is set as an objective and is advanced through including not only regional economic but also social and environmental criteria in annual business reports expressing achievements towards sustainability of agriculture in the region. As a practical approach, the Regionalwert AG has developed a set of 64 indicators, in 13 groups, to assess the sustainability of individual partner businesses.

Despite the rather limited scope of the initiative (in terms of partner businesses), a number of activities and events are considered signs of anchoring. The innovative concept of the Regionalwert AG and its success has raised broad media interest and has been covered in both scientific and public media, which may play an important role in the scaling up of the initiative. Furthermore, the founder has received several awards which further increase public visibility and credibility; he became an Ashoka fellow in 2009 and was nominated 'Social entrepreneur of 2011' by the Schwab Foundation, Boston Consulting Group and Financial Times Germany. As a consequence of increasing public interest in initiating similar concepts in other regions, the Regionalwert Trust was founded in September 2011.

The role of collaboration in transitions towards sustainability of agriculture

As the 'success' of an initiative, understood as the emergence of a transition at the regime level, can only be considered in retrospect, it is not yet possible to make definite statements about whether the three case studies referred to in this chapter have been 'successful'. However, the introduction and development of machinery rings in Scotland has resulted in incremental changes at the regime-level and assured them an established position within the agri-food regime. The other two initiatives have not (yet) advanced beyond the take-off phase (when a niche engages with the regime to initiate radical changes, see Darnhofer, this volume). However, the Regionalwert AG initiative, in particular, has demonstrated the potential to induce regime changes.

All three cases show that successful collaboration between the actors in each initiative is crucial for initiation and development. Moreover, joint analysis of cases suggests that collaboration takes a range of forms over time and at different stages of the development of the initiative and subsequently has different roles in the phases of potential transitions. This has, in return, variable effects on the initiatives themselves.

In the initial stages of niche establishment (and of potential transitions), interpersonal relations are important, in particular for the development of common concepts and strategies and the establishment of functioning structures, which was evident in the Regionalwert AG and CRIE Montado cases. At a later stage, the formalization of rules and structures can sometimes be a substitute for interpersonal relationships, as can be seen in the machinery rings case.

We could also verify the role of collaboration in the alignment between niche, regime and landscape to advance the transition, evident in the example of machinery rings. It was important in the pre-development phase of the transition (where CRIE Montado now stands), in which collaboration concerns mainly 'horizontal cooperation' and functions as a critical means for the niche establishment process – promoting innovation and actively experimenting and defining viable and coherent alternatives or solutions to the problems the regime is facing. It was also important in the take-off phase, where collaboration between niche and regime actors shape anchoring (and later mainstreaming) of the transition, concerning mostly 'vertical cooperation' processes. In this way, while vertical cooperation was important for anchoring the niche into the regime in the machinery rings and Regionalwert AG cases, horizontal cooperation seemed to be more relevant during initial phases. This was seen in CRIE where members expressed their need to increase it and delay vertical cooperation.

Factors affecting collaboration

To further understand how and which forms of collaboration have impacted on transition processes, it is necessary to understand the particular driving forces and barriers to collaboration in the three initiatives. The factors which influence cooperation can be grouped into those concerning actor relationships (roles, capacities, trust and commitment), objectives (expected benefits) and technical means, which are of course all interrelated.

Interpersonal and power relations are important, in particular, in the development of bonding and trust, envisioning common concepts and strategies, and establishing functioning structures. These were particularly evident in the Regionalwert AG and CRIE Montado cases. The latter case showed that asymmetries in social status can create asymmetries in power relations and communication, with consequences for collaboration processes. In the case of machinery rings, it was found that many farmers lacked a sense of ownership of the ring, despite the fact that they are de facto owners of the cooperative. However, the case study also revealed that this does not necessarily lead to failure of the initiative, as members perceived other benefits for themselves. Also, due to the nature of machinery rings, interesting power dynamics between members were observed. For example, collective buying power is a benefit for members who purchase commodities within the ring (e.g. farmers), which results in prices and income being driven down for other members (e.g. fuel suppliers). Commitment to collaboration was another key factor, in particular, concerning leadership: a committed personality with strong local networks and strong visions was a key factor for the success of the Regionalwert AG, while the continuously changing leadership did not support the development of CRIE Montado. In terms of formalized structures, both the machinery rings and Regionalwert AG cases have shown that management structures have a central role in facilitating collaboration between partner businesses (and between businesses and shareholders). Furthermore, they are essential in communication with regime actors. The fact that machinery rings have become involved in national level consultations and bodies can be interpreted as a sign of anchoring at regime level. In a very different way, niche-regime interaction is evident in the Regionalwert AG; the substantial public attention, increased visibility and a number of awards have made the group part of national level discourses.

Learning has played an important role in all of the initiatives and is relevant for the success or failure of the transition process. Two learning aspects are particularly important: learning regarding organizational functions and processes within initiatives; and learning on aspects related to agricultural management and sustainability. Machinery rings and Regionalwert AG have both institutionalized learning at different scales. Machinery rings has acted as a knowledge provider through advisory services and consultancies. Regionalwert AG has established an entrepreneurs' forum for knowledge exchange among partner businesses and the Regionalwert AG Trust. The Trust's objective is to provide knowledge and support to other regions aiming to set up similar initiatives.

The second group of factors concerns the objectives set by initiatives, and how these are pursued. Innovative forms of collaboration have been successful, particularly if they are a response to the day-to-day needs of members; often, stakeholders engage in collective action to reduce their feelings of isolation. These individuals, therefore, feel that they have gained something from collective action, even if these benefits are different to those originally envisaged by the initiative. Examples of this include the new links established between shareholders and organic business owners in the case of Regionalwert AG, or a direct economic benefit to actors as can be seen in the machinery rings case, where formalization of cooperation between farmers in resources and labour sharing is the novel aspect. However, it is not only utilitarian interests that are drivers in establishing new forms of collaboration. In CRIE Montado and Regionalwert AG, the explicit interests in environmental (and social) aspects of the collaboration have contributed to the development of the initiatives.

The third aspect concerns the means used to facilitate collaboration. The case studies show that these can evolve over time and do not necessarily have to be introduced at the outset to contribute to the successful establishment of initiatives. They are also the result of the aforementioned learning processes. Technical innovations such as computer software or direct debit technology, and service orientation (such as convenient opening hours and ease of payments) introduced by the machinery rings have clearly increased the attractiveness of machinery rings for farmers. Opposed to that is the lack of logistical infrastructure, which has been identified as a limiting factor in the physical and material collaboration between farmers and other agricultural actors in the Regionalwert AG.

It should also be noted that in all three cases, policies, particularly those aimed at fostering collaboration such as measure 124 (EAFDR 2007), have not played a significant role either in initiation or in the establishment of the initiatives. On the contrary, the lack of policies or political action in particular areas which are perceived as problematic for the sustainability of agriculture has been seen as the main driver for the establishment of a niche. In relation to CRIE Montado, the lack of policies focusing on integrated/ multifunctional approaches, and in the Regionalwert AG case, farm succession and increasing external capital in farms, were the core drivers for the formation of the initiatives.

Impacts of collaboration on regional sustainability of agriculture

As CRIE Montado and Regionalwert AG are relatively recent examples of collaboration, it is only possible to estimate the potential impact that they will have on regional sustainability. Only one of the three initiatives – the Regionalwert AG – is overtly pursuing greater regional sustainability in agriculture. However, the two other cases were important to study as they, at least partly, pursue objectives that meet criteria for regional sustainability.

CRIE Montado has contributed to the sustainability of agriculture by: (i) reinforcing the adoption of environmental actions by members in their own farm businesses (often linked to organic farming but also to new activities which are orientated towards sustainability, such as eco-camping or educational projects about sustainable farming); (ii) following and implementing the proposition of Jan Huijgen's multifunctional farms (Postema and Ziel, 2008), which calls for a re-invention of urban-countryside relations, sustainable farming, community development and a more humane, creative and balanced territorial rural development through knowledge exchange and lobbying; and (iii) developing and implementing a novel notion of multifunctionality with regard to the Montado system, thereby providing a new vision for regional sustainability with a system that has been increasingly recognized as traditional "well-adapted economically viable multiple-use agroecosystems for promoting sustainable modern development in many farming areas of the Mediterranean Basin" (Blondel, 2006:725). The Montado system has shown remarkable stability and sustained productivity over a period of 800 years or longer, and is characterized by high biodiversity. CRIE Montado's potential contribution to the regional sustainability of agriculture lies not only in the ecological but also in the social and institutional dimensions (regarding the informal social institutions evolving as the group's collaboration becomes more established). However, the initiative has had only limited impacts to date.

The contribution by machinery rings to sustainability of agriculture in the north-east and Borders regions of Scotland is primarily an economic and institutional contribution; it provides a mechanism that helps to sustain the economic viability of farms through reduced costs and greater efficiencies in terms of access to agricultural inputs across these regions, whilst at the same time facilitating agricultural 'management' at a regional level. Farmer collaboration through machinery rings provides a number of opportunities, including: (i) resource sharing, which helps farmers to reduce their fixed costs and to reduce over-capitalization and business risk in the agricultural sector; (ii) labour services, whereby additional income can be generated through labour supply (by individuals or farms), additional labour can be easily accessed (demanded) and employment opportunities are provided through the ring; (iii) trading of commodities, which helps farms to access benefits associated with economies of scale – which is particularly important for smaller farmers; and (iv) training accessible to members. Although as a mechanism, machinery rings are fundamentally based on collaboration and generation of social capital across the regions, the consequences for social sustainability are mixed – including positive and negative effects on agricultural labour. For example, rings provide a means for young farmers to find work and gain experience in the sector; but at the same time, rings facilitate lower retained labour on farms, as farmers can access the ring's labour pool on demand. Whereas machinery rings currently make an important contribution towards the economic and social aspects of the regional sustainability of agriculture, their impacts on environmental sustainability are mostly limited. However, there are specific examples where contributions are currently being made (for example in the development of renewable energy production) and opportunities have been identified for machinery rings to coordinate smaller scale collaborations for the purposes of implementing local or landscape-scale policies relating to the environment, therefore also potentially making a greater contribution in the future in that regard.

The Regionalwert AG Freiburg initiative contributes to the three dimensions of regional sustainability of agriculture (ecological, economic and social) in several ways.

Most important for the success of the initiative (and the most innovative aspect) is the creation of new linkages between (regional) shareholders who invest in organic farms and other businesses in the value chain within this region, thus providing investment capital for start-ups and supporting new entrants and young farmers as well as other businesses. This new link between the agri-food sector and financial markets is a key economic contribution to the sustainability of agriculture. With specific regard to collaboration, the initiative strengthens the regional organic agriculture value chain by supporting different forms of collaboration between farmers and business partners (including knowledge transfer mechanisms) that alter the structure of regional value-added chains towards increased integration. Beyond this, the initiative also contributes to sustainability by introducing the principle that supported businesses are not only valued by their individual economic performance but also by social and environmental criteria which contribute to regional sustainability. The benefits generated by the Regionalwert AG, by providing transparent investment options for citizens who support regional agricultural sustainability, should not only be regarded as (regional) economic returns (to shareholders), but also as social and environmental gains which benefit not only shareholders but the wider regional population.

The incumbent agri-food regimes discussed in this chapter all face farm succession issues to varying degrees. In all three cases, the contribution to social sustainability made by the initiatives has so far been either limited or contradictory (as exemplified with regard to labour supplied through machinery rings). This is the case even though the initiatives may potentially facilitate the establishment of young farmers and new entrants, as in the case of CRIE Montado and machinery rings; or if young farmers and new entrants are particularly targeted by the initiative, as in the case of Regionalwert AG.

The initiatives discussed here have highlighted several aspects which influence the situation of young farmers and new entrants at the regime level: (i) the size of farms (large in the case of CRIE Montado and increasing in size in Regionalwert AG) limiting the options for expansion or acquisition by new entrants; (ii) conventional capital markets which are not easily accessible for both groups; (iii) labour intensity which prevents young farmers from becoming involved in associations; (iv) the background and expectations of young farmers' partners and families, which may not prepare them for rural living; (v) lack of advanced training offered for young farmers or for new entrants with diverse educational and professional backgrounds; and (vi) the type of farming systems pursued by young farmers and new entrants. Despite general concern with farm succession and with inclusion of young farmers and new entrants, none of the three studies found policy measures in place effectively supporting these groups. Although these initiatives were open to young farmers and new entrants to varying degrees, the two groups did not play an explicit or particularly important role in the alignment of processes or regarding the initial innovations proposed in any of them.

Conclusion

It has been pointed out that collaboration is a central form of human agency and thus builds the foundation of the development of initiatives. One of the aims of this chapter has been to identify how innovative collaborations have influenced processes of transition towards regional sustainability. The socio-technical innovations discussed above include: a multi-level collaborative model working towards increased multifunctionality in the case of CRIE

Montado; the formalization of collaborative processes between farmers and novel technical solutions to facilitate resource-sharing in the case of machinery rings; and the direct financial involvement of shareholders in regional organic farms and businesses for the generation of economic, environmental and social returns in the case of Regionalwert AG. The way the forms and roles of collaboration vary with different stages of (potential) transitions was also discussed, as well as the contribution that these initiatives have made to the regional sustainability of agriculture.

However, in all three cases it is not only these novel forms of collaboration, but also other more conventional interactions and collaborations between actors, which play a significant role in the development of the initiatives and in their contribution to sustainability. Thus, initiatives can only successfully develop if the novel forms of collaboration (e.g. actor networks or objectives pursued) are combined with well-established forms of collaboration (e.g. management structures).

Moreover, regarding the effects of collaboration varying at different stages of the transition process, in CRIE Montado, collaboration seems to have increased relational connectedness and trust, and created the space to initiate and advance innovations during niche establishment, whereas interpersonal relationships have been substituted, to some extent, in the case of machinery rings by institutional facilitation. Thus, collaboration provides a medium in which the transition process is continuously shaped and advanced by actors, in interaction with pressures and opportunities from the landscape – a medium which is the key for innovations and for their development and mainstreaming. This highlights the importance of investing in social capital and supporting collaboration. Nevertheless, in all three cases there is a lack of support given to collaborative structures and processes by public policies.

This study has shown that different types of collaboration may have potential value at different stages of transitions; the role of collaboration varying according to the stage of transition reached. The returns obtained from supporting collaboration are, therefore, uncertain but risk is inevitable in developing innovative solutions, and such support will have a multiplying effect on the efforts of actors involved in transition processes.

References

Blondel, J. (2006) The 'design' of Mediterranean landscapes: A millennial story of humans and ecological systems during the historic period. *Human Ecology* 34, 713–729.

Darnhofer, I. (2015) Socio-technical transitions in farming. Key concepts. In: Sutherland, L-A., Darnhofer, I., Wilson, G.A. and Zagata, L. (eds) *Transition Pathways towards Sustainability in Agriculture: Case Studies from Europe*. CABI, Wallingford, UK, pp. 17–32.

EAFDR (2007) *European Agriculture Fund for Rural Development (2007-2013)*. European Commission. Available at: http://ec.europa.eu/agriculture/

Fuchs-Heinritz, W. and Barlösius, E. (1994) *Lexikon zur Soziologie*. Westdeutscher Verlag Opladen, Springer, Dordrecht, Germany.

Klischat, U., Klischat, U. and Habermann, I. (2001) Erfolgsbestimmende faktoren landwirtschaftlicher kooperationen aus sicht von betroffenen. In: Schwerdtle, J.G. (ed.) *Sammelband zum Symposium der Edmund Rehwinkel-stiftung: Vol. 15*. Betriebsgesellschaften in der Landwirtschaft. Chancen und Grenzen im Strukturwandel, Frankfurt am Main, Germany.

Peneva, M., Draganova, M., Gonzalez, C., Diaz, M. and Mishev, P. (2015) High nature value farming: Environmental practices for rural sustainability. In: Sutherland, L-A., Darnhofer, I.,

Wilson, G.A. and Zagata, L. (eds) *Transition Pathways towards Sustainability in Agriculture: Case Studies from Europe.* CABI, Wallingford, UK, pp. 97–112.

Pinto-Correia, T. (1993) Threatened landscape in Alentejo, Portugal: The 'Montado' and other 'agro-silvo-pastoral' systems. *Landscape and Urban Planning* 24, 43–48.

Postema, S. and van der Ziel, T. (2008) *Cityside Oasis: Or How to Bridge the Gap Between City and Countryside.* Eemlandhoeve, the Netherlands.

Roep, D. and Wiskerke, J. (2004) Reflecting on novelty production and niche management in agriculture. In: Wiskerke, J. and van der Ploeg, J.D. (eds) *Seeds of Transition. Essays on Novelty Production, Niches and Regimes in Agriculture.* Van Gorcum, Assen, the Netherlands, pp. 341–356.

Surová, D. and Pinto-Correia, T. (2008) Landscape preferences in the cork oak Montado region of Alentejo, southern Portugal: Searching for valuable landscape characteristics for different user groups. *Landscape Research* 33, 311–330.

Sutherland, L-A., Wilson, G. A. and Zagata, L. (2015) Introduction. In: Sutherland, L-A., Darnhofer, I., Wilson, G.A. and Zagata, L. (eds) *Transition Pathways towards Sustainability in Agriculture: Case Studies from Europe.* CABI, Wallingford, UK, pp. 1–16.

Chapter 7

High nature value farming: environmental practices for rural sustainability

M. Peneva[1], M. Draganova[2], C. Gonzalez[3], M. Diaz[4], P. Mishev[1]

[1]*University of National and World Economy, Sofia (Peneva_mm@yahoo.co.uk);* [2] *Institute for the Study of Societies and Knowledge, Sofia;* [3]*University of Évora;* [4]*AGROCAMPUS OUEST, Rennes*

Introduction

This chapter focuses on possible transition pathways towards sustainable agriculture through an understanding of the environmental, social, cultural and economic advantages of 'high nature value farming' (HNVF) and its implications for European regions. The overall objective is to explore how the implementation of various traditional agricultural practices aimed at nature protection and biodiversity conservation in high nature value (HNV) areas (primarily Natura 2000 sites) may lead to regional sustainability of agriculture and rural areas. HNVF is seen not only as an environmental solution but also as having a broader impact on the economic and social sustainability of agriculture and rural development at the regional level.

After the Second World War, rapid change towards productivist agriculture began to emerge in Europe and beyond. Productivist agriculture has been characterized as large-scale, intensive, industrially-based and expansionist, strongly supported by the state and primarily oriented towards increased productivity (Lowe *et al.*, 1993; Wilson, 2001; Burton, 2004). These changes were economically driven and established the current dominant productivist agricultural regime. However, productivist agriculture came with considerable associated environmental costs and negative impacts on rural regions. These included: overproduction, environmental pollution, biodiversity decline and landscape change, a reduction in farming jobs, rural exodus and the abandonment of many rural areas.

By concentrating production (economically and spatially), these trends have also created a polarized distribution of protection – and production-oriented landscapes in rural areas – the latter often having significant impact on the environment, on local rural economies, and on food quality and safety. Furthermore, this has led to a loss of much of the social meaning associated with farming and farmers in rural communities. A need emerged for greater understanding of the circumstances of peripheral rural areas, which required: (i) new approaches and innovations; (ii) new roles for farming; and (iii) new farming identities. At the same time consumer awareness regarding food quality and safety grew, together with rising demand for agricultural commodities in emerging markets and the inclusion, within European Union (EU) rural development and food quality policies, of traditional agri-food products. There is general consensus amongst stakeholders about the new role of agriculture and the need to shift towards more sustainable development of industry and rural areas (Brinks and de Kool, 2006).

This new role has been reflected in changes in European agricultural policy over the last two decades, which have been broadly characterized as a shift from productivism to post-productivism (Ward *et al.*, 2008). The academic literature provides differing interpretations of this development; there is no commonly agreed definition of post-productivist food regimes and their driving factors are difficult to conceptualize (Wilson, 2001). According to Gundelach (2005), the use of the term ‘post-productivism’ seems to relate to a “combination of the emergence of the EU CAP and a general (ideological) wish for alternatives to industrialised farming” (Gundelach, 2005:248). The transition to a post-productivist agricultural regime is also defined “as a struggle between agricultural and environmental interests, with the central axis of an environmental critique of modern intensive farming” (Ward *et al.*, 2008:119). Wilson and Rigg (2003) claim that the commonalities between different visions for the transition from productivism to post-productivism are the exogenous drivers of agricultural change. They highlight six interconnected indicators of post-productivism: policy change; organic farming; counter-urbanization; the inclusion of environmental non-governmental organizations (NGOs) at the core of policy-making; the consumption of the countryside; and on-farm diversification activities.

As discussed in Sutherland *et.al.*, and Darnhofer (both this volume), transition theory sheds light on sustainability as an ongoing process: it is a normative goal and demonstrates potential for radical change, either in production systems, policies, or in consumer behaviour and community awareness, depending on regional characteristics including institutions, regulations, negotiations, social norms, values and consents. Therefore, sustainability in farming is not only a goal to be achieved but also a process (Buttel, 2006; Marques *et al.*, 2012; Darnhofer, this volume). In line with this notion, transitions towards HNVF are approached in this chapter as an ongoing interactive process balancing three dimensions: economic, social and environmental.

The HNVF concept has the potential to become a key component of the European model of agriculture through its post-productivist and multifunctional characteristics, its contribution to the diversity of rural areas and through the conservation of biodiversity. HNV areas are acknowledged in the formulation of the new post-2013 CAP: they appear as part of the greening requirements as well as within new rural development tools focusing on environmental quality. HNVF has been developed across Europe and has shown promising results in terms of the environment and for farm economics, with significant reductions in levels of inputs (Brinks and de Kool, 2006).

In this chapter, HNVF is approached from the multi-level perspective (MLP) of transition theory (Darnhofer, this volume) and in light of the notion of multifunctional agriculture, despite the broad and divergent approaches to the latter in academic and policy debates (Wilson, 2008). The three main concepts of HNVF, MLP and multifunctionality are outlined in the second section. Transition processes have been analysed at the regional level in three European countries, through three case study areas in Bulgaria, France and Portugal. The third section has two functions. First, we describe the three case studies which form the chapter's empirical focus. Second, we elaborate on the adaptation of farming practices as transition and as anchoring processes which occurred in three dimensions: technological, social and institutional. We conclude by discussing the transition as an ongoing process where innovations developed within the three HNVF initiatives lay the basis for potential radical change.

The research is largely based on qualitative methods of data collection. In the three case studies, semi-structured interviews were conducted with representatives of various stakeholders: farmers, including young farmers and new entrants; local and regional authorities (mayors and municipality representatives); representatives of public–private bodies (chambers of agriculture); agricultural officers and experts; non-governmental organizations (environmental NGOs and others); entrepreneurs (hotel-keepers, food processors); and other key informants. Desk-based research was also carried out to provide the data for analysis (at the meso and local levels) of the contextual features of the case studies, including historical, socio-economic, cultural and agricultural characteristics. A review of relevant policy documents (EU and national) relating to nature conservation and biodiversity issues was also carried out.

The 'high nature value farming' concept

The broad concept

The HNVF concept, which first emerged in 1993, recognizes the relationship between certain types of farming activity and 'natural values' (Baldock *et al.*, 1993), and that the conservation of biodiversity in Europe depends on the continuation of low intensity, low input farming systems across large areas of countryside (Baldock *et al.*, 1993; Bignal and McCracken, 1996; Opperman *et al.*, 2012). HNVF is an extensive system and conforms to Natura 2000 regulations and other requirements covering protected areas. In general, it implies grassland management with benefits for biodiversity conservation and habitat protection. Its cornerstone is semi-natural pastures, meadows and orchards, as well as peripheral semi-natural features such as large hedges and copses (Opperman *et al.*, 2012).

HNVF is carried out in all European countries. It is an integral part of the landscape, covering a broad range of landscape types and having common features across Europe. Farm operations are based on management systems that use regional breeds of livestock. Local skills, which complement the climate and geography of the locality are drawn upon; the use of artificial fertilizers and chemicals is minimized; small scale cultivation often takes place alongside livestock production; and largely uneconomic and declining labour-intensive management practices, such as shepherding, are often revived[1].

HNVF is a top-down approach, which is emphasized in EU directives for the protection of HNV areas[2]. These encourage bottom-up initiatives which adopt environmentally-friendly land management practices for HNV grassland, biodiversity protection and nature conservation in HNV areas in Europe[3]. They include: 'optimising the

[1] WWF Danube-Carpathian Programme, project 'High Nature Value Farmlands – Recognizing the European Importance of SEE landscapes', www.panda.org/dcpo

[2] Birds Directive 79/409 and Habitats Directive 92/43

[3] A European Commission report (2009) indicated that the increase in agricultural land, climate change, tourism, and poor land management leads to loss of biodiversity in the 27 EU countries, and governments must take measures for its protection. This informed the need for sustainable management of HNV farmlands through HNVF. It designates broad categories of low-intensity farming and livestock breeding that use environment-

use of available measures under the reformed CAP, notably to prevent intensification or abandonment of HNV farmland, woodland and forest and supporting their restoration'[4]. The challenge is how to discourage HNV farmers from abandoning or intensifying their farming system in the search for better returns or for survival as farmers – both paths are major causes of biodiversity loss (Beaufoy and Marsden, 2011).

HNVF in the context of the EU multifunctional agriculture paradigm

The concept of the multifunctionality of agriculture has been discussed over the past three decades in the context of numerous research and policy debates. A review of the literature reveals diverse viewpoints and the evolution of the concept. In 1988 the Commission of the European Communities emphasized the notion of multifunctionality by recognizing the multiple functions of agriculture and more broadly, of rural areas. Agriculture and rural areas are considered an important part of territorial economic and social cohesion, environmental protection and conservation, and for the vitality of rural communities. More recently, a broader definition of multifunctionality encompasses and emphasizes the generation of non-commodity outputs including environmental goods, food safety, animal and plant health, animal welfare standards and quality of life in rural areas (Council of the European Union, 1997).

From within academia, Wilson (2001) argues that the debate about productivist and post-productivist agricultural regimes may not reflect the "diversity, non-linearity and spatial heterogeneity that can currently be observed in modern agriculture and rural society" (Wilson, 2001:96). Instead, he has proposed the notion of a multifunctional agricultural regime which goes 'beyond post-productivism': "a regime that conceptually, temporally and spatially follows on from the post-productivist *transition*" (ibid:92,95). This notion also emerged from the potential of agriculture and agricultural land to fulfil diverse and new functions within society. These include a production function (non-commodity goods and services), an ecological function, a cultural function and a recreation function (Knickel and Renting, 2000; Jongeneel *et al.*, 2005; van Huylenbroeck *et al.*, 2007). These changes, and new societal demands, imply not only a change in policy but also a change in the institutional environment (Jongeneel *et al.*, 2005), creating new rules, norms, values and attitudes.

We approach HNV areas as multifunctional rural areas where protection and production functions coexist and are both increasingly relevant, in that they combine different socio-economic, environmental and cultural functions in one and the same region. European rural landscapes have been shaped by humans through pluri-activity and multifunctional modes of rural occupancy over millennia. In general, HNVF is connected with the conservation of rural communities and rural lifestyles, large numbers of family/small farms, stronger local economies, regional food security, food quality and safety, and the amenity value of the landscape (Ilbery and Bowler, 1998).

---

friendly land management practices for HNV grassland, biodiversity protection and nature conservation in HNV areas in Europe.

[4] COM(2006) 216 Final Communication from the Commission halting the loss of biodiversity by 2010 – and beyond. Sustaining ecosystem services for human well-being.

HNVF from the multi-level perspective

In the MLP, the niche is considered the locus of radical innovations. The niche is created by actors at the local level and initiates various innovations that change the dominant (incumbent) regime (see Darnhofer, this volume). Therefore, we explored the interaction of HNVF initiatives (in space and time) with the dominant regime and sub-regimes.

In the transition process, the socio-technical landscape pressures the regime to alter practices where persistent problems exist, thereby creating windows of opportunity that might be used by niches. In the case of HNVF, we observed the strong and direct influence of the socio-technical landscape on niche development. Support instruments at national and/or regional levels associated with HNVF classification can be interpreted as one of the pressures at the regime level, which is strongly connected to the trends in EU CAP instruments and reforms towards multifunctionality and post-productivism, as described in previous sections. Therefore, the HNVF approach is considered (mostly) as a top-down approach and thus the socio-technical landscape creates 'windows of opportunity' for the niche to 'break through' the regime and to encourage niche-regime interaction. To some extent, this challenges the theory of transition, the foundation of which is that the transition has to happen through bottom-up initiatives. Despite being a top-down approach, HNVF inspires bottom-up initiatives[5] such as adopting environmentally-friendly land and landscape management practices for HNV grassland, biodiversity protection and nature conservation in HNV areas in Europe, whilst at the same time promoting rural development.

Regional context of HNVF and transition

The case studies

The areas studied are the Bessaparski Hills in southern Bulgaria, the Saint Amarin Valley in eastern France, and the Baixo Alentejo in southern Portugal (Fig. 7.1). Case studies in each region varied in terms of size of area, actors involved, time span and sectors concerned. The Bulgarian initiative focuses on an individual protected area (Natura 2000 zone). The French case focuses on a valley in a Regional Nature Park (RNP) in the Vosges Mountains. The Portuguese initiative focuses on three municipalities – Mértola, Barrancos and Almodôvar – within which there are several protected areas (including Natura 2000). In Bulgaria the initiative began following the implementation of a project by the Bulgarian Society for the Protection of Birds (BSPB) aimed at conserving important grasslands and encouraging farmers to adopt land management practices that conserve and maintain existing biodiversity and habitats. The French initiative is undertaken by new entrants in farming to bring abandoned farmland back into use and to conserve agri-pastoral areas in the St Amarin Valley. The Portuguese initiative started with the project 'Valuing Mediterranean wild resources' coordinated by the Municipality of Almodôvar and promoted by the Association for Mértola Heritage Protection (ADPM). A key driving force was the Programme for the Economic Value of Endogenous Resources (PROVERE), a

[5] It was mentioned that the French case is a bottom-up initiative but it was possible after the CAP reform in 1992.

horizontal policy tool aiming, in this case, to increase the territorial competitiveness of the region by promoting sustainable exploitation of local natural resources. This was to be achieved through the promotion of traditional local non-wood forest products as a way of associating nature conservation and a system of production, which develops the region, and by promoting the amenities and multiple-use of the Montado extensive land use system. The range of actors involved in the conception and evolution of the three initiatives varies considerably. In Bulgaria it consists of farmers, BSPB and representatives of the local offices of the Ministry of Agriculture and Advisory Services. In France new entrants and, later, established farmers, authorities at the municipality and community of municipalities levels, RNP officers, Chamber of Agriculture and civil society participated. In Portugal, a collaboration within the PROVERE project involving a total of 97 partners (2009), including producers exploring non-wood forest products, producers' associations, local and regional NGOs, local action groups (LAGs), enterprises and business associations, municipalities, conservation institutions, research institutions and universities were involved.

The time span of the three initiatives also varies, as the French case is the most longstanding (having started in the 1980s). This enabled it to run through the four phases of the transition process proposed by Rotmans and Loorbach (2010). These include: predevelopment of the niche, with the setting up of new entrants on municipal land with the support of mayors; the development of networking at the local and regional levels, establishing an experimental (specific) policy instrument after the reform of the CAP (1992) (take-off phase); and the widespread adoption of agri-environmental policy (acceleration phase). However, the last phase (stabilization) was not completed because the environmental problem has largely been resolved and the local network has therefore become weak and in some respects unstable. The other two initiatives are more recent: the HNVF initiative in Bulgaria started in 2008; and in Portugal the preparation process started in 2007 following the approval of the PROVERE project but the official start was not until 2009. Therefore, it is difficult to discern any long-term impacts of the last two initiatives and the comparison within this chapter is focused on the take-off phase for all three initiatives.

In the three case studies, a key policy instrument was agri-environmental measures. The distinction between the actors involved and time scale of the initiatives determines differences in the design, implementation and diverse impacts on the regional and local level of the measures, as well as the approach of the transition process. In the Saint Amarin Valley, the transition was a bottom-up process where adoption and implementation of agri-ecological practices led to the designation of the territory as an HNV area. In the Bessaparski Hills, the specific requirements and natural constraints were imposed through national environmental legislation and policy, as top-down restrictions. In Bulgaria, agri-environmental measures concerning HNV grasslands were designed at the national level, and were not focused exclusively on, or adapted to, the area of the Bessaparski Hills. This contrasts with the French case, where measures regarding the improvement of pastures were developed at regional level and implementation was delegated to local mayors. With regard to the Portuguese municipalities, both bottom-up and top-down initiatives took place: the NGO (ADPM) was heavily involved in creating a shared strategic vision for the region, one which paid attention to economic, social and environmental sustainability. However, the notion of rural territorial development based on endogenous resources, and the policy and funding schemes structuring it, was a top-down measure as part of the PROVERE programme, which was then adopted by the NGO and other partners.

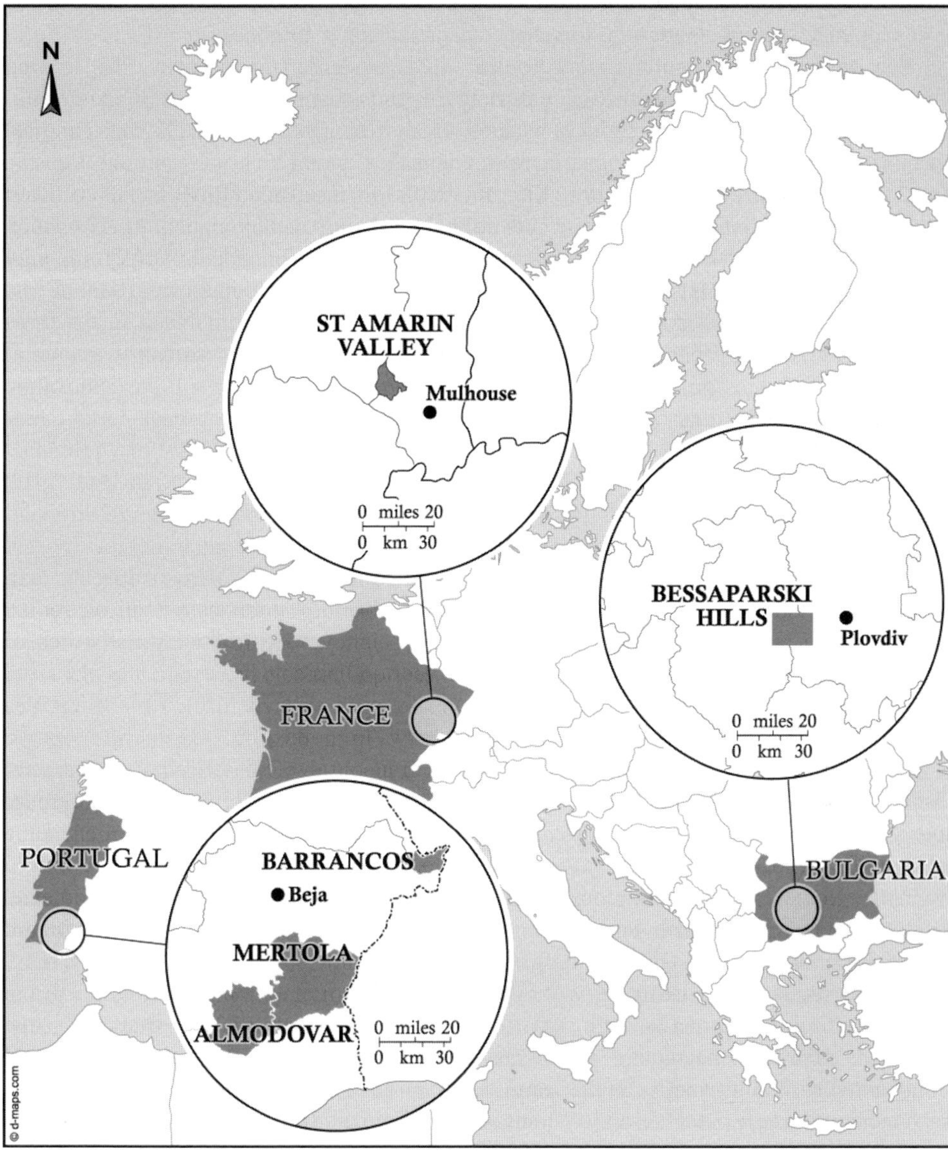

Fig. 7.1. Location of the three case studies: Bessaparski Hills, Saint Amarin Valley and Baixo Alentejo (Source: authors).

Nevertheless, for the success of the transition process and creating radical change, the key point of the three initiatives is their common goal. This common goal is the long-term conservation of HNV landscapes and the maintenance of existing biodiversity and habitats, protecting the quality image of local products, and stimulating sustainable development in rural areas by creating livelihoods for farmers and other local producers, multi-level collaboration, partnership and territorial competitiveness. These are the means by which the initiatives aim to generate new sources of income for local people and contribute to the local economy, thereby supporting social and sustainable development.

Several common economic and social trends underlie these aims. The leading economic sectors of the study areas are agriculture and food processing in Bulgaria; non-food industry, commerce and services in France; and the service sector in Portugal. The organization of agricultural business, farm size and farm structure also varies between the three initiatives, as well as within the initiatives themselves. However, a common contextual characteristic of these three case study areas is an ageing population. Therefore, the participation of young farmers and new entrants (in the French initiative) are very important both for the vitality of the territory and for the implementation of HNVF, its associated measures and practices. The proportion of young people involved in the three initiatives increased over time as they appear to have become more sensitive to issues of environmental protection. Young people were also more attracted to the innovative proposals, could adapt more readily to new ideas and were more willing to undertake new initiatives and risks.

The anchoring process

Elzen *et al.* (2012) introduced the term 'anchoring' for emerging forms of linking, and proposed that anchoring can occur in each of the three dimensions (technological, social, and institutional) (see Darnhofer, this volume). The data show that all these forms of anchoring have emerged in the case studies.

Social anchoring – networking

The MLP views transition as a non-linear and multi-actor process. In the case studies, we observed various stakeholders/actors enmeshed at all levels in an interconnected, dynamic and co-evolved manner. These actors were acting and interacting through different production systems with institutions, regulations, negotiations, social norms, values and consents. In the French and Portuguese cases, some of the stakeholders involved were powerful regime actors.

Collaboration and networking were the most widespread forms of interaction in the initiatives and a similar networking context was observed in all three case studies. On the one hand, environmental actors and local officials sought solutions to conserve local biodiversity and landscapes. On the other hand, in each region some farmers/producers implemented farming practices and systems which allowed the conservation of local natural resources and landscapes, often 'innovating with the old'. As a result, new relationships between various actors with diverse interests have been built at different levels: local, regional and national. In the French case, the first collaboration between actors took place at the local level between new entrants looking for farmland and local mayors who were struggling to prevent their municipal land from becoming fallow or developing into forest. When agri-environmental measures were introduced (in the early 1990s) at the local level, the network extended into the inter-municipal level thanks to local hybrid actors (a young farmer, an agronomist/landscape manager who was also an elected local official, and two civil servants in administrative bodies) who helped overcome some resistance at the local and regional levels. Together, actors (local and regional, public and private) built a programme for local development based on landscape management, re-emphasized the value of local farm products and created an agri-environmental measure to support landscape management.

In the Portuguese initiative a collective efficiency strategy, itself a multi-level collaboration process, developed a public–private partnership and networks across the Alentejo region. Similarly, it benefited from collaboration between local and regional actors who came together to progressively develop a local strategy and vision for the region. However, from the beginning this initiative was implemented as a horizontal policy tool. The initiative materialized through the elaboration of a strategic vision for development, implemented through integrated development plans which included an action programme and the establishment of partnerships. ADPM used the opportunity to develop the collective efficiency strategy in a participatory process that involved different public and private promoters (including farmers and also municipalities and private enterprises). ADPM, which is more pro-active than the farmers, was the key initiator and remains the main coordinator. Nonetheless, farmers' engagement has increased and some farmers also belong to ADPM – thus they are both beneficiaries and participants in a process of empowerment.

In the Bulgarian case, the initiative was led by a national environmental NGO (BSPB). The group promotes collaboration between farmers through training, encouraging farmers' awareness of the value of practising HNVF. Collaboration is recognized as very important, but in practice it has negative connotations with the past when cooperation was subject to, and controlled by, the state. Farmers in the region maintain good contact and exchange ideas, information and experience with each other, mediated by the BSPB. There are informal networks of HNV farmers in the region but they also retain contact and exchange visits with farmers from other regions of the country, facilitated by the BSPB. Good relations and collaboration were also established between HNV farmers and representatives of the Regional Office of the Agriculture Advisory Service (ROAAS).

In all three case studies, networking processes facilitated anchoring between niche and regime actors and other domains. Networking also generated collaboration between actors and created social links leading to the development of social capital. In the Bulgarian case, for example, despite negative connotations, informal networks have developed between farmers throughout the region and beyond. In the French case, an association has been created to implement an action plan for the management of collective equipment, training sessions, and creation of a farm shop to promote local products. In the Portuguese case, the collaboration process was strengthened from the start by a participatory process developed to produce an efficiency strategy. In all three cases, the transition to HNVF involved the establishment of a social component to implement collaborative projects, the impact of which, in terms of social links and economic opportunities for local actors, potentially outweighs the original environmental objectives.

The success of innovating and adapting agricultural practices to natural conditions depends on the involvement of actors and stakeholders from local and regional levels being embedded in formal and informal organizations, and/or other bodies and networks. It is important, therefore, to highlight the differences across the case studies in the processes of collaboration. In the French and Portuguese case studies there was greater cooperation amongst farmers, which was supported by several state or regional bodies/institutions. Bulgarian farmers were less collaborative and less open to dialogue over management practices. In the latter case the involvement of strong personalities was essential for activities such as cooperation or exchange of information. Collaboration and interaction also resulted in devolving decision-making processes to the local level (France); meetings, exchange visits, training and cooperation under the guidance of an environmental NGO (Bulgaria); and multi-level communication and power imbalances between institutions (Portugal).

Institutional anchoring – new rules, values and beliefs

Institutions as 'rules of the game' structure human interaction and activity (North, 2005). Transitions to sustainability in agriculture may not be primarily technology-driven but are likely to involve elements such as social innovation and require changes in beliefs and values by all societal actors (Darnhofer, this volume). The innovations related to changes in institutions concern changes in actors' values, beliefs and interpretation of rules (Smith, 2007), and changes in normative institutions' (regulations, policies) formal and informal rules. Common features in social innovation across the three cases include raised awareness and new understanding between farmers, locals, NGOs and representatives of the regime regarding the value of nature conservation. These values subsequently led to the development of new environmental attitudes and behaviour.

Empirical evidence from the research shows that increased awareness and new values emerged from the bottom up in the French case, where new entrants encouraged others wishing to enter farming by re-working abandoned land. This was also the case in Portugal where producers became aware of the value of traditional products and of the benefits of protected areas. Conversely, in the Bulgarian case new values came from the top down through the process of collaboration with BSPB; farmers adopted attitudes which valued indigenous biodiversity. New knowledge and skills about HNVF practices increased the awareness and responsibility of farmers involved in the initiative.

In respect of the changes in normative institutions (policies and governance), in all three cases changes in agri-environmental policy at the national level (through different programmes and measures) directly affected the development of initiatives. For example, in the Bulgarian case, the implementation of agri-environmental measures at regional or local levels may have been influenced by changes in institutional arrangements, depending on the specific characteristics of the protected areas, and such changes may have modified the concept of these measures.

In the anchoring process between the niche and regime, 'hybrid actors' can appear who are engaged at both regime and niche levels. Hybrid actors were identified only in the French case study and included the association, technical networks and the Chamber of Agriculture. As in the French case, hybrid centres can bring together local representatives, civil society actors and farmers.

Learning, knowledge, skills and capacity building

In the MLP, learning should focus on how networks are built and maintained, and on the complex interactions between technical and institutional aspects linked to innovation (Darnhofer, this volume). Learning and improved skills and knowledge are notable outcomes from the case study initiatives. *Training* was the most popular form of learning found in the three cases. In France, training in landscape management brought together farmers and local representatives. In Bulgaria, collaboration with the BSPB started with training sessions to raise farmers' awareness of the value of HNVF. As beneficiaries of advisory services, farmers maintained good relations with their regional office. In Portugal, the training process was usually organized by key promoters of the collective efficiency strategy, took the form of visits to production units, and often covered the production and processing of wild resources – with an emphasis on the differentiation and diversification of products by innovating and improving traditional products.

Learning about *marketing activities* concerned direct marketing of local food (part of the HNVF concept); diversification; promotion of traditional foods to new customers; creating better quality food; and increasing farmers' incomes. New skills were required, such as good hygiene practices; design skills (jar choice, packing, and labelling); and marketing techniques to retain customers and to build long-term relationships.

Learning with regard to the organization of *collective marketing* resulted in the opening of a regular open air market in the St Amarin Valley and in the collective management of a farm shop. In addition to production and processing, *capacity building* included marketing products and access to consultation services. At this level, there was promotion of entrepreneurship and related skills, as well as investment in searching for innovative marketing channels, such as in the European 'eco' market. Capacity building in the areas of production, processing and marketing was also connected to events aimed at supporting networking among producers. It should be noted that professional networking and trust-building received particular attention, and it was possible to verify learning regarding: (i) the importance of collaboration and partnership in achieving economic scale; (ii) the importance of sharing leadership and commitment; and (iii) the synergetic effect of sharing knowledge and visions for the region. Furthermore in Bulgaria, consultation and training on the implementation of agri-environmental measures, organization of visits to exchange experiences, and provision of information, technical advice and practical tools, were major constituents of the learning process. *Advisory services* and sharing knowledge about sector-specific legislation and funding, as well as promoting know-how on licensing processes or funding schemes enabled by institutions, also contributed to the learning process in the Portuguese and Bulgarian cases. In Bulgaria in the field of financial issues and grant application procedures, the regional agricultural advisory service was a key actor; whereas at the level of marketing design and good hygiene, it was the BSPB Mobile Advisory Centre who played the key role and made an important contribution to the provision of advice. In all three cases the empowerment of individual producers was notable.

Discussion

What changed?

Changes in terms of understanding, policy framing, activities, collaboration and networking, and values have occurred in all three initiatives but at different levels and across different time-scales. In general, the three initiatives changed local culture, particularly for farmers living in protected and less-productive areas, such that these areas have been re-valued. The effect was a greater concern and awareness for the protection of local ecosystems, resources and inhabitants.

A very important change in the three initiatives was the strengthened process of collaboration and networking. Partnerships built new bridges between actors with different interests, which resulted in collective action and boosted innovation. Strong horizontal networking was created in the French and Portuguese initiatives, integrating multi-level actors (from local to regional and even national in the French case). The networks facilitated exchange of information, knowledge and experience; built skills to help actors participate in decision-making processes; and increased trust and confidence.

During the transition new bridges were created between the agri-food regime and sub-regimes through activities such as recreation (France) and rural tourism (Bulgaria). In the Portuguese case, no new bridges were created at this level as Mediterranean Wild Resources traditionally had links to agri-food and forestry regimes, and with amenities provision (for example tourism and trekking) and biodiversity conservation. Subsequently, in adapting farming systems to multifunctionality, an important institutional anchorage has occurred: farmers' identities were challenged and changed. French farmers involved in the transition remained food producers but also shifted to being landscape managers; Portuguese producers also remained producers but seemed to perceive themselves more as agricultural entrepreneurs and/or service providers exploring high quality innovative products and services. Bulgarian farmers remained as commodity producers revealing, nevertheless, a trend towards increased enterprise and a shift towards becoming service providers.

Moreover, the comparison shows that there are common features in all three initiatives that counter trends embedded in the dominant regimes. All three challenged the homogenization of production structures and of products, through diversification and multifunctional agricultural activities. Intensification was reduced (or at worst not increased) and the proportion of young people involved, or interested, increased counter to the trend of ageing in the rural and farming population.

All of the above changes have reinforced the transition towards multifunctionality of HNVF in the three areas, related to conservation of biodiversity and landscape quality combined with the need to sustain livelihoods and diversify agricultural activities. These changes created potential opportunities to secure new sources of income and employment but in a balanced, place-based model of territorial management in which landscape and local communities and their ways of life were prioritized.

Following this search for place-based solutions for balanced territorial development and new meanings of farming, another key factor which encourages the transition to HNVF is public demand for traditional quality food products. This leads to the evolution of marketing behaviour and activities, and the use of local certification schemes. The initiatives facilitate mechanisms through which farmers initiate direct sales, shortening the producer–consumer chain. Farmers become more flexible and have choice in planning and realizing their sales, seeking the best conditions and reducing their dependence on intermediaries. Currently, the three initiatives have succeeded in increasing local knowledge and know-how, reinvigorating awareness of quality local products and enabling responses to new consumer demand.

Furthermore, in the three case study areas, the development of HNVF and quality product marketing has been combined with local heritage and biodiversity protection through rural tourism activities. Diversification to such activities (e.g. gastronomy, nature-use activities) expands the promotion of regional/local heritage and contributes to the multifunctional characteristics of these rural areas.

Conclusion

One of the key lessons learned from comparing the three initiatives was the crucial role of normative institutions and funding opportunities, identified as key drivers in the emergence and development of transitions. In the MLP, successful transition of niche innovation occurs when the socio-technical landscape opens up a window of opportunity by putting pressure on the regime. In other words, the successful development of HNVF innovation would lead to transition utilizing policy support from one side, as financial payments, and from the other as restrictions imposed on the regime (such as encouraging intensive farmers to become engaged in sustainable agriculture). The CAP regulations (landscape pressures) shape the current legal framework for promotion, adaptation and further development of the initiatives. But the practical implementation of HNV rules and the relevant agri-environmental measures vary between the cases. In France, it is shaped as a bottom-up approach where policy with regards to such measures has been efficiently implemented; whereas in Bulgaria and Portugal, the top-down approach is followed as HNVF policy is adopted and implemented by the measures and included in their National Rural Development Programmes (2007–2013). This process is more recent, policy driven (provides a window of opportunity forcing the regime to recognize the niche) and in both cases is reinforced by bottom-up processes driven by actors like NGOs. These NGOs can be active promoters of policy requirements and facilitators of interrelations between niche and regime actors (for example between farmers and administrators). These examples illustrate how national policies can be adhered to and adequately implemented at regional level, taking into account particular regional characteristics and needs.

Another important lesson learned concerns the role of producers in collaboration and networking processes and their potential effect on the pathway to transition. In the Bulgarian case, farmers were considered to be beneficiaries of extension services and public policy; they did not take part in creating the local development plan and this lack of participation could affect further adoption of the innovation (Callon *et al.*, 2001). In the French case, farmers were empowered and officially commissioned to manage a public good; they became the leaders of the initiative and could influence public decisions and policy making. In the Portuguese case, they were part of the collaboration process in the same way as other actors involved in the collective efficiency strategy. It is perhaps too early to draw definitive conclusions about the success of the initiatives but we can suppose that this approach is efficient in terms of adoption of the innovation and management of public funding. Based on the research we cannot conclude that a bottom-up initiative makes the process easier.

However, we can verify that transition is an ongoing process. The transition process reached the take-off phase only in the French initiative, as it was the most longstanding. It is too early to conclude that the initiatives led to a transition in the Bulgarian and Portuguese cases, as they are at an early stage of transition. From the MLP perspective, for transition to happen there should be alignment at the three levels: niche, regime and landscape. In our cases the alignment of the three levels was achieved in the French case but it has not yet been achieved in the Bulgarian and Portuguese cases due to the time-frame and other constraints. The research on the transition to HNVF raises questions as to whether the transition has an end, and what follows the stabilization of the transition.

Finally, transitions towards multifunctionality of HNVF are reflected in the three cases in respect of: (i) a search for place-based solutions to balanced territorial development (with lesser polarized distributions of protection- and production-oriented landscapes in rural areas, and lesser impacts from the latter on environment, local economies and communities); and (ii) through a search for new conceptions of farming and farming identities in rural areas. Some success in achieving these objectives could be observed, particularly in the French case, thereby confirming that HNVF has the potential to become a key concept in the European model of agriculture through its post-productivist and multifunctional characteristics, its contribution to the diversity and sustainability of rural areas, and through the conservation of biodiversity.

References

Baldock, D., Beaufoy, G., Bennett, G. and Clark, J. (1993) *Nature Conservation and New Directions in the EC Common Agricultural Policy*. Institute for European Environmental Policy, London, UK.

Beaufoy, G. and Marsden, K. (2011) *CAP Reform 2013 Last Chance to Stop the Decline of Europe's High Nature Value farming*. Available from http://www.efncp.org/download/

Bignal, E.M. and McCracken, D.I. (1996) Low-intensity farming systems in the conservation of the countryside. *Journal of Applied Ecology* 33, 413–424.

Brinks, H. and de Kool, S. (2006) Farming with future: Implementation of sustainable agriculture through a network of stakeholders. In: Langeveld, H. and Röling, N. (eds) *Changing European Farming Systems for a Better Future. New Visions for Rural Areas*. Wageningen Academic Publishers, Wageningen, the Netherlands, pp. 299–304.

Burton, R.J.F. (2004) Seeing through the 'good farmer's' eyes: Towards developing an understanding of the social symbolic value of 'productivist' behaviour. *Sociologia Ruralis* 44, 195–215.

Buttel, F. (2006) Sustaining the unsustainable: Agro-food systems and environment in the modern world. In: Cloke, P., Marsden, T. and Mooney, P. (eds) *Handbook of Rural Studies*. SAGE Publications Ltd, London, UK.

Callon, M., Lascoumes, P. and Barthe, Y. (2001) *Agir dans un Monde Incertain: Essai sur la Démocratie Technique*. Editions du Seuil, Paris, France.

Commission of the European Communities (1988) *COM (88) 601. The Future of Rural Society*. CEC, Brussels, Belgium.

Council of the European Union (1997) *Council – Agriculture*. Press Release: 343 – Nr.12241/97, Brussels, Belgium.

European Commission (2009) *Report on Article 17 of the Habitats Directive: Conservation Status of Habitats and Species of Community Interest (2001-2006)*. Available from: http://ec.europa.eu/environment/nature/knowledge/

Darnhofer, I. (2015) Socio-technical transitions in farming. Key concepts. In: Sutherland, L-A., Darnhofer, I., Wilson, G.A. and Zagata, L. (eds) *Transition Pathways towards Sustainability in Agriculture: Case Studies from Europe*. CABI, Wallingford, UK, pp. 17–32.

Elzen, B., van Mierlo, B. and Leeuwis, C. (2012) Anchoring of innovations: Assessing Dutch efforts to harvest energy from glasshouses. *Environmental Innovation and Societal Transitions* 5, 1-18.

Gundelach, P. (2005) Visions in agriculture. *Sociologia Ruralis* 45, 245–262.

Ilbery, B. and Bowler, I. (1998) From agricultural production to post-productivism. In: Ilbery, B. (ed.) *The Geography of Rural Change*. Pearson Education Ltd., Harlow, UK, pp. 57–84.

Jongeneel, R., Polman, N. and Slangen, L. (2005) Why are farmers going multifunctional? In: *XIth International Congress of the EAAE. The Future of Rural Europe in the Global Agri-Food System*. Copenhagen, Denmark. Available from: http://library.wur.nl/file/wurpubs/

Knickel, K. and Renting, H. (2000) Methodological and conceptual issues in the study of multifunctionality and rural development. *Sociologia Ruralis* 40, 512–528.

Lowe, P., Murdoch, J., Marsden, T., Munton, R. and Flynn, A. (1993) Regulating the new rural spaces: The uneven development of land. *Journal of Rural Studies* 9, 205–222.

Marques, F., van der Ploeg, J.D. and Dal Soglio, F. (2012) New identities, new commitments: Something is lacking between niche and regime. In: Barbier, M. and Elzen, B. (eds) *System Innovations, Knowledge Regimes and Design Practices towards Transitions for Sustainable Agriculture*. INRA - Science for Action and Development, pp. 23–46.

North, D. (2005) *Understanding the Process of Economic Change*. Princeton University Press, Princeton, New Jersey.

Opperman, R., Beaufoy, G. and Jones, G. (eds) (2012) *High Nature Value Farming in Europe. 35 European Countries - Experiences and Perspectives*. Verlag regionalkultur, Heidelberg, Germany.

Rotmans, J. and Loorbach, D. (2010) Towards a better understanding of transitions and their governance. A systemic and reflexive approach. In: Grin, J., Rotmans, J. and Schot, J. (eds) *Transitions to Sustainable Development. New Directions in the Study of Long Term Transformative Change*. Routledge, New York, pp. 103–220.

Smith, A. (2007) Translating sustainabilities between green niches and socio-technical regimes. *Technology Analysis & Strategic Management* 19, 427–450.

Sutherland, L-A., Wilson, G. A. and Zagata, L. (2015) Introduction. In: Sutherland, L-A., Darnhofer, I., Wilson, G.A. and Zagata, L. (eds) *Transition Pathways towards Sustainability in Agriculture: Case Studies from Europe*. CABI, Wallingford, UK, pp. 1–16.

van Huylenbroeck, G.V., Vandermeulen, E. and Mettepenningen, A. (2007) Multifunctionality of Agriculture: A Review of Definitions, Evidence and Instruments. In: *Living Reviews in Landscape Research* 1 2007, 3. Available at: http://www.livingreviews.org/lrlr-2007-3

Ward, N., Jackson, P., Russell, P. and Wilkinson, K. (2008) Productivism, post-productivism and European agricultural reform: The case of sugar. *Sociologia Ruralis* 48, 118–132.

Wilson, G.A. (2001) From productivism to post-productivism . . . and back again? Exploring the (un)changed natural and mental landscapes of European agriculture. *Transactions of the Institute of British Geographers* 26, 77–102.

Wilson, G.A. (2008) From ‘weak’ to ‘strong’ multifunctionality: Conceptualising farm-level multifunctional transitional pathways. *Journal of Rural Studies* 24, 367–383.

Wilson, G.A. and Rigg, J. (2003) ‘Post-productivist’ agricultural regimes and the South: Discordant concepts? *Progress in Human Geography* 27, 681–707.

Chapter 8

Transition processes and natural resource management

G. Vlahos[1], S. Schiller[2]

[1]*Agricultural University of Athens (gvlahos@aua.gr);* [2]*Institute for Rural Development Research, Frankfurt am Main*

Introduction

Natural resource management is one of the main challenges faced when designing policies aimed at increasing sustainability in rural areas, not only because of the complexity of natural processes but also due to complex interdependencies, spatial and temporal scale differences, and multiple policy dimensions involving a wide range of stakeholders. Although natural resource management in rural areas has been a matter of discussion for over a century, debates continue about the most appropriate trajectories towards sustainable management of natural resources in the countryside (Wilson, 2007).

The case studies analysed in this chapter will focus on several resource regimes, including management of water resources, agri-food and tourism. High water quality is considered as a public good, co-produced during the agricultural production process, along with private goods and services (Cooper *et al.*, 2010). There is a tendency in the socio-technical transition literature towards the analysis of private goods and market-based solutions, which is not always relevant when public or quasi-public goods are concerned. For this reason, we draw upon an institutional economics perspective to complement the theoretical framework elaborated earlier (see Darnhofer, this volume).

The first characteristic that should be stressed is that when a common good, in our case a natural resource, is concerned, in the majority of cases collective decision-making is preferable since it reduces transaction costs and increases efficiency compared to individual bargaining procedures (Vatn, 2005, 2010). Collective means of civic coordination have commonly been criticized as costly in terms of time and resources. However, when the effectiveness and efficiency of economic instruments are assessed, it is assumed that all necessary institutional arrangements are in place, which may not be the case (Anderies *et al.*, 2004; Foxon, 2010), constituting an additional cost which, if taken in account, could alter the overall performance of these instruments. Furthermore, collective decision-making might be slower but could provide the time and space to foster learning and the capacity to maintain the ecological functions of the resources in question – in this case resilience (Vatn, 2005). On the other hand, institutional settings have a significant influence on trade-offs and manoeuvres within the negotiation process, as well as on the enrolment of actors (Callon, 1986; Arce and Long, 1992). In addition, social interaction in resource management encourages political participation (Adger *et al.*, 2005), creating an interface between politics, economy and society. According to Vatn (2005) there are three main aspects of a resource regime: access to and distribution of the resource; the related transaction costs; and the perceptions, interests and values arising from the interaction of actors and agents. Concerning alternative policy approaches in order to resolve the problem of environmental impacts created by human activity, the same author suggests that public intervention could be directed towards either regulating inputs or the production process as a whole. The former is a somewhat imprecise method, especially when diffuse pollution is encountered as in the cases we analyse here, or in terms of emissions which usually raise transaction costs to prohibitive levels.

Concluding this short introduction, it should be noted that the above approach does not follow the neo-classical approach in which a common resource is defined as one from which nobody may be excluded. In this chapter, a common natural resource is considered a 'good' serving society. In that sense, this approach grants rights and responsibilities to citizens to make choices concerning future developments (Vatn, 2005).

Setting the scene

The case study areas

The common characteristic of the three case studies (Fig. 8.1) is that they focus on areas where agriculture is exerting pressure on water resources. The French study region is located in the north of Brittany and includes the catchment areas of five rivers running into Lannion Bay. Apart from a wooded valley, the majority of the catchment is agricultural, arable land, complex farming systems, and agricultural land interspersed with natural areas. Multiple compounding factors, such as specific photoperiods, high water temperatures, a low degree of sea water turbulence, and relatively high levels of nutrients provided by converging streams, contribute to the development of green algae which periodically washes up on the beaches of the flat, semi-enclosed and shallow Lannion Bay.

The water catchments in the Mangfall Valley, the German study region, are situated 50 km from Munich in the district of Miesbach, Upper Bavaria, where one third of the district's total area is used for agriculture. The region is notable for hosting the largest continuous area under organic farming in Germany, due to the organic farming programme supported by the water supplier of the city of Munich. Water protection, partly overlaid by protected landscape areas, has been designated in the head areas of drinking water wells.

The Greek study area is the department of Imathia in the region of Central Macedonia, northern Greece. Since 2006 it has constituted part of a Nitrate Vulnerable Zone as the area is part of the overall water catchment of two rivers. Furthermore, the delta of the two rivers is part of a NATURA 2000 protected site. Irrigated crops account for 93% of the area, a far higher average than the national figure of approximately 30%. Furthermore, it was the region with the highest rate of fertilizer application according to the most recent available data (Beopoulos, 1996).

For the research, interviews with 56 individuals representing institutions, local and regional governments and authorities, researchers, agricultural cooperatives, farmers, consultants and experts were conducted (24 for Lannion Bay, 10 for Mangfall Valley and

22 for Imathia). The interviews were complemented by the analysis of publications produced by key actors and also by the analysis of secondary literature such as scientific reports, articles, action plans and newspaper articles.

Socio-technical landscape pressures

The main economic landscape pressure identified was the global trend towards agricultural intensification. This trend appears to have had an influence in all three areas. In Lannion Bay it was experienced as a shift towards corn as the main feed stuff; in Mangfall as one of the two main options for small-scale, part time farmers; while in the case of Imathia, initially the expression of intensification was towards an increase of production in order to achieve higher subsidies. Intensification can include increased utilization of inputs (agri-chemicals and water resources); mechanization of production; development of annual crops to the detriment of pastures and the decline of wildlife areas (including important biotopes) (EEA, 2010); and increased product specialization. Several impacts of this process have been identified. Apart from the most prominent, such as water and soil quality deterioration, loss of wildlife and agricultural biodiversity and landscape degradation, important socio-economic impacts have also been observed such as increased dependence on providers of inputs, usually resulting in higher costs for farmers. A further consequence of intensification was the marginalization of farms and significant areas of farmland. Agricultural land not appropriate for intensification due to climatic, soil or geographic limitations has also been abandoned. Smaller households not able to cope with the intensification process tend to abandon farming, especially in EU-designated Less Favoured Areas (LFAs) like Mangfall, whilst larger or specialized farms concentrate their efforts in the most productive areas.

The widespread notion of sustainability 'institutionalized' through the Rio conference in 1992, and its particular implications for the agricultural sector, represents another pressure from the socio-technical landscape, the main element of which is an increased concern about negative externalities caused by intensive agriculture. This concern has been expressed in different ways in the different case study areas. For example, in the case of Lannion Bay it took the form of continuous media and legal pressure, especially when acute problems arose like the death of a horse in the bay. In Mangfall, the role of agriculture in ensuring clean water for the district of Munich has been recognized in official planning documents. In Greece as a whole, and Imathia in particular, spatial and temporal alienation of urban populations from their rural origins led to a de-legitimization of support for rural areas, especially through environmentally harmful payments for withdrawal of excess production. Closely related to the sustainability discourse described above has been an increased awareness of food safety which has heavily influenced two of the case study initiatives. The demand for organic food, one of the opportunities presented in the Mangfall initiative, can be partly attributed to this trend. Public calls for stricter regulations on pesticide use could be considered to be the regional manifestation of the food safety issue in Imathia.

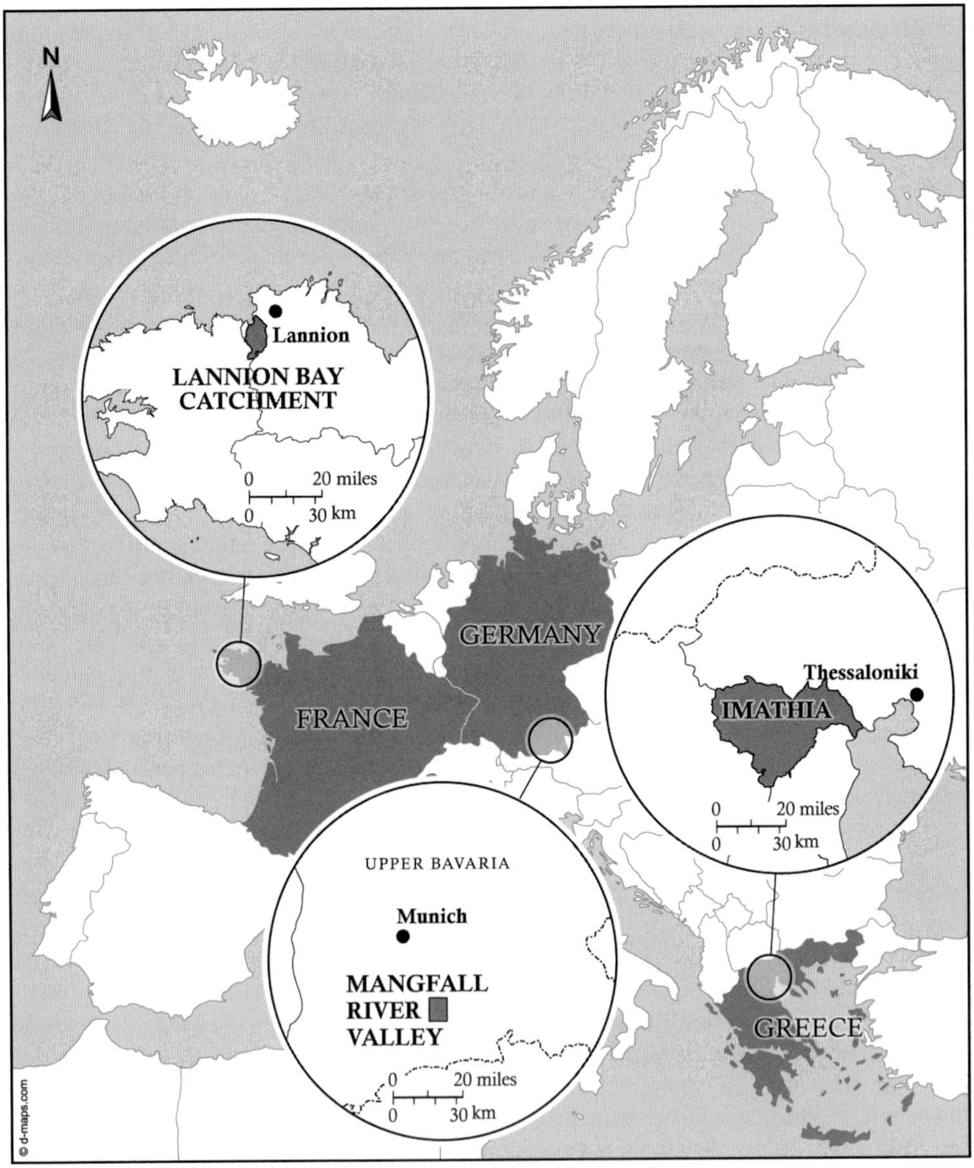

Fig. 8.1. Location of the three case studies: Lannion Bay, Mangfall Valley and Imathia (Source: authors).

The social, economic and environmental pressures noted above have generated responses from both authorities and wider society. Their most prominent manifestations are policy measures launched by the EU, state and/or regional and local authorities which have opened windows of opportunity for niche development.

Description of the incumbent regimes

The pre-transition period in the three case study areas was characterized by the interdependency of agri-food regimes and water management. However, the differences between regimes are manifest mainly in two dimensions: the technological and the institutional. Thus, while in the German case agriculture is of an extensive nature, typical for a Less Favoured Area (LFA), and contributing to a rural landscape of high value with links to tourism and amenity provision, in the other two areas intensive agricultural practices prevail, exerting pressure on water resources. Furthermore, within the intensive systems the Greek subsidy-oriented production system may be distinguished from the French, mainly market-oriented system. Another difference between the German case and the other two study areas was that a public utility service, Stadtwerke München GmbH (SWM), established for the management of water resources, played the role of the principal agent, while no special arrangements for water management existed in the other two.

In the German study region agriculture, seen as a land use regime, faces the challenges and opportunities of a LFA of considerable scenic beauty, in the vicinity of a metropolitan urban centre. A special cultural landscape has developed over the centuries as a result of the topography, the small size of farms, cultivation practices in permanent grassland and dairy farming. This type of cultural landscape, typical of mountainous LFAs, could be considered as a sub-regime, as it provides amenities for recreation and tourism. Small farm holdings in general face increasing economic pressures when structural adjustments in agriculture are slower in LFAs. These pressures result in high levels of uncertainty amongst farmers about whether to continue farming. Investing in the farm and intensifying production lies at one end of the spectrum of potential responses; at the other is abandonment of farming. However, the existence of successors in the area suggests that organic farming provides a viable alternative to the dilemma. The requirements regarding animal welfare and feeding in organic agriculture have been increasingly strict in the past few years and derogations such as permission to use small amounts of conventional feed if no organic feed was available, have been gradually phased out. Tie-stalls will be prohibited in Bavaria from 2013. Consequently, a number of small dairy farmers have been obliged to reorganize their installations for cattle breeding, requiring costly investment. The role of organic advisory services is relatively prominent in the process of reorganization. A number of organic farming associations, particularly Naturland consultants, are active in the region. However, the difference between organic farmers in the Mangfall Valley and the other case study areas is their spatial concentration and their motivation for adopting organic agriculture.

The initiation point in the Greek study area was the early 1980s when Greece entered the EEC and a single function regime based on intensive fruit production began. Under the Common Market Organisation for Fruit and Vegetables, subsidies were given for withholding large quantities of produce from the market as a means to stabilize prices. The main aim of farmers and their collective organizations thus became the exploitation of land in order to produce maximum quantities, and thereby increase the revenue gained through subsidies. Consequently, quality was not a predominant concern. In terms of the important technical dimensions of the regime in Imathia, the production process was conventional, characterized by the production of fresh fruit, almost exclusively peaches, as well as peach production for canning, which gradually marginalized other crops. For the latter, Greece held, and still holds, an important segment of the global market. Paradoxically though, links between peach producer groups and local markets, and development of rural tourism initiatives, are virtually non-existent in contrast to the gradual integration of the local wine industry.

The human-societal dimensions of the regimes could be seen as determined by the CAP, alienating farmers from the market. It is significant that 50 producer groups were created in order to withdraw production. The use of new technology contributed towards the intensification of production and a broad range of related agents and actors were formed. Informal provision of information and advice by agri-chemical suppliers and equipment providers eclipsed the role of public extension officers (Koutsouris, 1999). Dependence on subsidies also influenced the formation of clientelistic relations through direct partisan linkages between the heads of farming cooperatives, central government and party leaderships bypassing local administrations and, in many cases, disrupting linkages with local society.

The agri-food regime in Lannion Bay, has been differentiated into two sub-regimes: food production and food consumption. The food production sub-regime was key as it was affected by changes imposed in order to mitigate algal proliferation ('marées vertes' or 'green tides'), which appeared even with an average N concentration of 30 mg/l. Nitrogen in river water is the only factor that can be readily influenced to limit algal growth rates (Ménesguen, 2003). The technical solution proposed recently has been to increase the area of grassland and decrease the extent of cropland in the territory, in order to limit nitrate losses. To achieve this collective goal, farmers were encouraged to decrease areas of corn and increase grassland, with a collective target of 60% grassland in the Utilised Agricultural Area (UAA) of the watershed before 2015. This measure introduced more complexity into local farm systems, creating problems in terms of winter fodder provision and the reduction of dairy production (LTA and BV-LDG, 2010).

The local food system is based on long supply chains and agri-industrial processing of dairy products which is well established in Brittany. Cooperatives and private firms are commercially powerful and are able to influence farmers' decision-making through business opportunities. Green Algae Plan measures appear to conflict with the current dynamics of downstream industries, which have not developed a strategy around local production techniques and quality since their preferences lie with large volumes, standardized products and exports, although tourism has been a mainstay of the local economy in Lannion Bay. The proliferation of green tides causes considerable inconvenience for this sector, not only in terms of physical, visual and olfactory impacts but also in terms of health. The decomposition of algae on the beaches creates H_2S vapour, a gas which is highly toxic to humans and other terrestrial animals (INERIS, 2009). An estimation of decline in property values in the littoral lies between 30% and 50% and the number of summer tourists has dropped by about 20% (SCE, 1999).

The public utilities service SWM provides the institutional setting for the supply of potable water in Mangfall. SWM and the Municipality of Munich are unpopular with many farmers because they have been buying land within the catchment since 1880. In 1911, around 1,164 ha had been bought whilst in 2011, this figure had more than doubled to around 2,700 ha. In 1950, a village located in an area later designated as a water protection area was purchased, the inhabitants resettled and their houses demolished. Intensification of agriculture in the region resulted in an increase of nitrogen concentration in potable water. As a response, in 1992 the SWM founded the Organic Farming Support Programme and secured the Organic Farming Association as a partner to fulfil consultation and administrative requirements. In 2001 an application to extend the water protection area created conflict between the SWM and inhabitants (including farmers) in the water catchment. Catchment municipalities were afraid of the rising cost of public buildings and restrictions, or even prohibition, on construction; whilst farmers were concerned about restrictions in farming. Restrictions have already been imposed on farming in the existing water protection areas, particularly on SWM owned fields which are leased by farmers, who are therefore obliged to comply. Even on private land farmers often feel that their management decisions are hampered.

The transition processes

Emergence of niches

The emergence of niches in the three case studies followed both bottom-up and top-down trajectories. A bottom-up process in the French case, initiated by a small group of people sharing similar views, political orientation and interest in farming went through several phases enrolling, over time, networks and actors at different levels. In the two other cases, hybridization of regime actors was necessary in order to instigate change. A public service provider (SWM) in the city of Munich directly contracted individual farmers in the Mangfall Valley in order to safeguard provision of clean potable water while in Imathia AGROCERT, a certification organization under state management, provided farmer groups with the means to overcome a serious obstacle and compelled existing local networks and actors to refocus their strategy.

Three phases describing human-societal and institutional aspects have been distinguished as having already taken place in Lannion Bay, while a fourth is expected. The predevelopment phase took place between 1980 and 1990 when some farmers located around the town of Corlay, leaders in grassland systems and also members of the minor farmers' union formed a group who met regularly to exchange ideas about practices and the economic performance of grassland systems (CETA). With the support of local and national left-wing politicians who gained power after the socialist victory in the presidential elections of 1981, they created an association of farmers (CEDAPA) to share information and to extend the association over a larger area. During that phase, although membership increased even in Lannion Bay, there was no collaboration between niche and regime actors and relationships were rather confrontational because of divergent points of view about agricultural development. The take-off phase took place during the 1990s. Participation was growing and the concept spread, mainly in the west of France. The leaders of CEDAPA formalized the specification of the farming system. At the same time, they continued to experiment and improve their technical knowledge and developed new collaborations with research institutes and the Ministry of Agriculture, to legitimize their systems. They also extended their participation in networks and committees, and established links with environmental NGOs resulting in increased influence at the regional level, although this was not felt locally.

The acceleration phase may be divided into two stages. The first stage saw a failed acceleration from 2003 to 2007. Although a new network was formed comprising alternative organizations, the collaboration between niche and regime actors stagnated and the local strategy was restricted to merely changing agricultural practices in order to reduce nitrogen loss. Finally, after 2008 it became clear to all parties that technical solutions were not sufficient to solve the problems, and hence elected local representatives and farmers from the Professional Committee of Agriculture Lannion Bay sought new solutions.

The major technical change proposed was the replacement of corn or cereals with grassland, with potential consequences for the entire production system. A collective objective was to have 60% of the area under grass by 2015 and, consequently, farmers had to learn how to manage the production of grass in order to secure enough fodder. In order to overcome this uncertainty, creation of a fodder bank was encouraged – supplied by farmers who had no need to feed cattle (such as pig producers) – and a plan was made to construct a collective dehydration unit. The second change was to reduce nitrate inputs and feedstuff imports from other regions, with the aim of reducing nitrate concentration in local water courses to 10 mg/l or less. The third change concerned the management of farm wetlands. Although experts on nitrogen flows in grassland systems had proposed the shift to fodder systems based on grass instead of corn, and the message had been supported by CEDAPA; the difference in values between the dominant regime and the niche seem to have been too large to bridge and the acceleration failed (Diaz *et al.*, 2013).

The organic farming initiative in the Mangfall Valley commenced in 1992, mainly to counter the decreasing quality of drinking water. Farmers received support for organic farming in the form of a payment for contribution to water quality. In addition, the SWM supported contract farmers to market their products. For small and extensive farm holdings in the district, organic farming seemed to be a good alternative which became even more attractive with SWM payments. Organic farmers were sometimes perceived as service providers to the SWM, a hybrid actor, since it constituted part of the regime and the niche, dictated the solution and formally contracted individual farmers (about 140 contracts). This conforms to the Pigovian principal-agent model approach (Vatn, 2005). Despite founding a group in the late 1990s to support their interests, farmers have so far failed to organize collectively. The conflict between SWM, the city of Munich and the Mangfall Valley appears to have been weakened as the SWM has bound farmers through long-term contracts and made supporters out of potential opponents (Ruhland, 2011). At the policy measures level, the initiative benefits from payments through agri-environmental measures of the EAFRD which, when added to SWM compensation, represents an attractive financial incentive. As a consequence, however, the price for leasing agricultural land has increased. Contractors of SWM aimed to expand their farms for two reasons. The first was to compensate for yield reduction (particularly common for organic farms) and the second was because payment levels depended on the extent of area farmed organically.

The niche in Imathia was based on adoption of the integrated farming standard 'AGRO2'. Although its main aim was to enhance environmentally friendly practices in agriculture, AGRO2 was initially used in order to address an important deficiency in the local agri-food regime; a failure to ensure an acceptably low level of pesticide residues. The procedure followed previously (post-harvest control of production) was somewhat dysfunctional and ineffective. Compliance required standardized procedures for all participants and for all aspects of the production process.

The role of advice was of key importance, as individual farmers tended to make decisions following consultation with experts. An informal advisory system was created with the involvement of several networks, including private management consultants assisting producer groups to set up and implement the management system, and private professionals offering technical expertise. The advisory system was supported by public sector experts (academics, researchers and administration staff) as well as previously neglected local research institutions. Finally, a network of private certification agencies was created and accredited by the state accreditation agency (AGROCERT).

The process of anchoring

In the initial phase in Lannion Bay, even a scientific report which suggested that reducing nitrate concentration was crucial in controlling green tides (Ménesguen, 2003) was questioned by regional mainstream farming organizations and some leaders of the agri-food industry. However, institutional anchoring at the local level did occur through the creation of a Local Professional Committee of Agriculture (CPA) bringing together farmers from both regime and niche (Diaz *et al.*, 2013). At the technical level, the issue of nitrate reduction from agriculture constituted a strong common reference point and enabled the connection of different networks. After consecutive technical solution failures, local elected representatives and farmers from CPA sought further scientific support. Experienced researchers endorsed the proposition of the niche network, acting as hybrid actors linking niche and regime networks. They also proposed collaboration with conventional farmers but were unable to secure funding until the death of a horse, which may have been poisoned by rotting algae in 2009, forced the Government to act.

Furthermore, farmers did not choose between two models but instead empirically combined elements from both, according to the characteristics of each farm. Ansaloni and Fouilleux (2006) considered that this selection of practice, called 'technical hybridization', consists of the appropriation of selected low inputs and also perceived cost practices by intensive farmers. Finally, a trend towards larger extensive farms was observed within the dominant regime, facilitated by the retirement of older farmers, although the niche tried to retain medium-sized units in order to enable young people to enter agriculture.

CEDAPA was initially very close to the political left (Déléage, 2004), as its principles were based on left-wing values. These historical roots remained strong, although the association gradually became more open. This may explain the difficulty faced by the niche in influencing the dominant regime. Farmers and/or their local representatives changed their perceptions regarding grassland systems during the ensuing 15 years. However, they continued to hold strong opinions about the early supporters of these systems. Under these circumstances, some farmers adopted the role of hybrid actors, facilitating anchoring by adopting only technical changes and enrolling farmers with whom they shared economic, but necessarily political, perspectives. Productivity, however, was not open to debate; a proposal for a decrease in the number of animals was abandoned following resistance by farmers.

In the Mangfall Valley catchment, the initiative focused on organic farming because this system was widely accepted as the most effective in protecting groundwater. Subsidies appear to have been a crucial factor encouraging adoption in the area, together with the fact that dairy agriculture in the Miesbach District already operated largely according to organic principles, hence most farmers could convert to organic farming without prohibitive investment. Nevertheless, the increased incidence of weeds in grassland observed on many contract holdings, as well as the use of expensive organic feedstuff for high milk-yielding cattle breeds (such as Fleckvieh), which were needed to prevent undernourishment in cattle, created decreasing yields and increased costs. There is evidence that some farmers could not cope well with organic farming practices (interview data). Consequently, organic farming gained a bad reputation among many farmers (including other organic farmers) in the region. As SWM, the principal actor in the initiative, was detached from the production process, farmers were not supported adequately in terms of technical issues. In fact, advisory services were provided by organic farming associations but these were not specifically focused on the initiative. The relationship between the principal agent and contractors has gradually evolved and improved over the years. Contract farmers organized themselves as an interest group and became an important partner for the SWM, increasing their negotiating power. At the last renewals of the support programme in 2010 and 2011, SWM proposed a contract of 1 year, as they anticipated extension of the water protection area. Eventually, the deal struck was for 15 years and at a higher premium per hectare (+22%), indexed to the cost of living.

The dairies collecting milk in the SWM catchments started establishing sales channels for organic dairy products and paid a premium for organic milk. In addition, an attempt was made to create a catchment farmers' market in the SWM headquarters but due to limited quantities, contract farmers ceased their participation in this market. The organic support programme seems to have reinforced the slowdown of structural change, hence dairy farming and the resulting distinct cultural landscape was conserved. Consequently SWM promoted the 'Munich Water Way', a bicycle tour through the water catchment area. However, conflict still exists in the process of expansion of the water protection area. Although contract farmers and the SWM are partners in water protection, all farmers are reportedly firmly against expansion of the water protection area, as they are concerned that they will not receive compensation.

The agri-food regime prevailing in Imathia up to the mid-1990s was not viable in economic terms. Change in the CAP, which was radical and abrupt unlike other policy shifts, found producer groups in the area facing a situation where export markets had either been lost or were maintained through disguised export subsidies. The latter appears to have been countered with technical barriers to trade, such as strict food safety regulations (for example, concerning pesticide residues) founded on increasing consumer awareness, and this resulted in the rejection of whole shipments of canned peaches in 1998/1999. The option to check all shipments before leaving the country was neither effective nor practical hence the urgent need to ensure that the final product could comply with restrictions in order to maintain export markets. In Imathia, AGROCERT's (a standardization and certification organization controlled by the Ministry and thus by definition a regime actor) hybridization towards the role of a niche instigator is the first evidence of institutional anchoring.

Local professionals (mainly agronomists) who, until then had been committed to input provision accompanied by the relevant technical advice, were the first to adopt innovations and adapt to changes. AGROCERT collaborated with the existing network of experts to enrol producer groups to the niche, unlike other attempts in Greece which focused on regime actors (Koutsouris, 2008). Thus, technical anchoring took place with the transformation of an existing well established network of private professionals (agronomists), from input providers to providers of advisory services. This was due in part to socio-technical landscape pressures which were gradually rendering their activities obsolete. First, changes in Common Market Organisation (CMO) rules concerning subsidies for withdrawal of production and then, almost immediately after, a new set (and uncertainty) of rules for pesticide 'Maximum Residual Limits' affected their businesses. The required shift from higher productivity to safe and quality products created the 'perfect' instrument for the operationalization of AGRO2. This standard had, as its primary aim, an environmentally friendly way of farming. The first part of the standard necessitated a series of procedures and monitoring which proved to be appropriate to achieve a tightly controlled process. New technological developments, mainly in the field of spatially explicit methodologies and environmentally friendly farming practices, have been trialled through EU-supported research projects jointly run by local institutes, producer groups and private consultants. Futhermore, organizational and management innovations, triggered by the implementation of the AGRO2 standard, have been supported by the EU. As increasing consumer awareness called for a continuous reduction of pesticide residues, the controlled and documented production process made rapid adaptation to consumer demand possible.

Although its main aim was to enhance environmentally friendly practices in agriculture, AGRO2 was initially used in order to address an important deficiency: the failure by the market to ensure an acceptably low level of residues in final production. Nevertheless, realization of the initiative's potential led to a re-orientation of producer group goals when implementing Integrated Crop Management (ICM), in order to reduce costs and improve quality. In the case of canned peaches, the whole ICM project can be seen as a counter-oligopsony measure. Cooperatives/producer groups that have participated in the niche have seen their negotiating power increase within the value chain; hence private merchants have actually been following the market trends set by the cooperatives.

A distinction can be made between the anchoring process in Lannion Bay and that in Imathia. In the first case, a bottom-up initiated niche expanded and strengthened, first horizontally among 'peers' and, when momentum had been gained, recruited and networked amongst other institutional strata. In the Imathia case, a hybrid actor enrolled the leadership of groups such as producer groups, transforming networks and creating linkages among collectives, networks and state agencies, and providing the space for negotiation. In the Mangfall Valley case, the public principal agent clearly imposed and enforced rules through formal individual contracts. All efforts to change this relationship seem to have failed and even farmers' attempts to cooperate involved trying to counteract the negotiating power of the SWM, rather than challenging the institutional setting.

Conclusion

Three interwoven regimes were observed in Lannion Bay: water resources, agri-food and tourism. Although there were many interdependencies, the ones that were manifest were the pressures exerted by agriculture and positive links between the three regimes have not yet been established. The fact that controversy remains over the causes of green tides has contributed to the debate on who should bear the cost of control measures. Simply focusing on technological change in a restricted bilateral negotiation between farming leaders and policy makers, and more or less disregarding the institutional setting, led to 10 years of ineffective attempts to solve the problem. This blatant policy failure could also be attributed to inherent limitations of agri-environmental measures which only covered income loss and additional costs. The allowance for a small incentive was probably not enough to stimulate the institutional change needed. Currently, a process of moving towards a collective solution to the problem of green tides appears to have begun but it is still in its initial stage. A shift of focus may be detected towards regulating the production process instead of inputs, now that institutions have been built, groups are engaged and linkages are slowly being established between them. This fact, and the cultivation of relationships with public agencies, local authorities and advisory services, has been an uphill struggle but could create space and allow time for collective learning and the strengthening of relations with civil society. The tensions however, resulting particularly from the timeline imposed by the French Government, remain. The result has been a return to decisions taken bilaterally by farming leaders and policy makers and the exclusion of niche actors and NGOs from the collective process.

In the Mangfall Valley, a market solution was sought for the resource regime. The right to pollute, in that case, lay clearly with local farmers, although a shift in that aspect of the regime can be seen through the gradual acquisition of land by the public authority and re-lease to farmers. This may explain the reluctance of farmers to accept expansion of the water protection area, as it also suggests a re-allocation of rights in favour of the public. The two regimes involved, those of land use and water resources, have been articulated in a system drawing from the Pigovian model of principal–agent. Thus, transaction costs that would have been incurred if the compensation was decided through individual bargains were significantly reduced. In order to achieve the desired level of water quality, the choice was taken to regulate land use and production practices. The process was facilitated by the existence of an established set of rules and a control system in the form of organic farming. In the same way, policy measures like agri-environmental payments and marketing initiatives were used, with differing degrees of success, to accelerate the transition process. Knowledge and information structures available to all organic farmers accommodated their respective needs. The market solution selected, however, allowed neither space for mutual learning, nor encouraged specific collaboration amongst farmers, agents, local authorities or research networks.

In Imathia, the regime studied was an agri-food regime. Its interdependencies with the water resource regime, although formally strong, were constrained by efforts to maintain another collectively managed good: the quality and image of the main product (the peach). The strategy chosen was to regulate the production process. In this case study, the high number of farms made coordination of individual efforts an almost impossible task. Hence, an institutional change was a prerequisite in order to successfully manage the necessary technological changes. A way of reducing costs has been to use a novel institution, proposed by a public actor: a standard designed to protect natural resources and biodiversity. In this case, however, there were no established information, auditing-control, or advisory structures as there had been in the Mangfall Valley. Hence, the public certification agency, producer groups, input providers, researchers and experts had to redefine their role and establish, expand and re-orientate networks, playing the hybrid role of both regime and niche actors. This collective coordination scheme imbued the agri-food regime with environmental elements, although the initial objective was not environmental protection. On the other hand, the process relieved farmers' organizations from clientelism, empowering them and strengthening their bonds with local society.

In all three case studies two regimes; agri-food and water resources, coexisted, with varying power imbalances, creating tensions and friction. At the same time, the initiatives studied were of different origin and directions, from a typical top-down process in the Mangfall Valley and Imathia, to a bottom-up process, gradually gaining space among peers and governance levels in Lannion Bay. In all of these seemingly contradictory cases, the multi-level perspective has provided a useful concept in our attempts to disentangle the transition process.

Acknowledgement

The authors are grateful to Marion Diaz for her invaluable contribution to this chapter.

References

Adger, W.N., Brown, K. and Tompkins, E. (2005) The political economy of cross-scale networks in resource co-management. *Ecology and Society* 10, 9.

Anderies, J., Janssen, M. and Ostrom, E. (2004) A framework to analyze the robustness of social – ecological systems from an institutional perspective. *Ecology and Society* 9, 18.

Ansaloni, M. and Fouilleux, E. (2006) Changement de pratiques agricoles. Acteurs et modalités d'hybridation technique des exploitations laitières bretonnes. *Economie rurale* 292, 3–17.

Arce, A. and Long, N. (1992) The dynamics of knowledge: Interfaces between bureaucrats and peasants. In: Long, N. and Long, A. (eds) *Battlefields of Knowledge: The Interlocking of Theory and Practice in Social Research and Development.* Routledge, London, UK, pp. 211–246.

Beopoulos, N. (1996) The impacts of agricultural activities. In: Papaspiliopoulos, S., Papayiannis, T. and Kouvelis, S. (eds) *The Environment in Greece 1991-1996.* Sideri, Athens, Greece (in Greek), pp. 141–177.

Callon, M. (1986) Eléments pour une sociologie de la traduction - La domestication des coquilles Saint-Jacques et des marins-pêcheurs dans la baie de Saint-Brieuc. *L'Année Sociologique* 36, 175–208.

Cooper, T., Hart, K. and Baldock, D. (2010) *The Provision of Public Goods through Agriculture in the European Union.* Report prepared for DG Agriculture and Rural Development, Contract No 30-CE-0233091/00-28. Institute for European Environmental Policy, London, UK.

Darnhofer, I. (2015) Socio-technical transitions in farming. Key concepts. In: Sutherland, L-A., Darnhofer, I., Wilson, G.A. and Zagata, L. (eds) *Transition Pathways towards Sustainability in Agriculture: Case Studies from Europe.* CABI, Wallingford, UK, pp. 17–32.

Déléage, E. (2004) *Paysans de la Parcelle à la Planète. Socio-antropologie du Réseau Agriculture Durable.* Syllepse, Paris, France.

Diaz, M., Darnhofer, I., Darrot, C. and Beuret, J-E. (2013) Green tides in Lannion Bay: What we can learn about niche-regime interaction? *Environmental Innovation and Societal Transitions* 8, 62–75.

European Environment Agency (EEA) (2010) *The European Environment. State and Outlook 2010.* Office for Official Publications of the European Communities.

Foxon, T. (2010) Managing the transition towards sustainable regimes. A co-evolutionary approach. In: Marletto, G. (ed.) *Creating a Sustainable Economy. An Institutional and Evolutionary Approach to Environmental Policy.* Routledge, Abingdon, UK, pp. 115–131.

INERIS (2009) Résultats de mesures ponctuelles des émissions d'hydrogène sulfuré et autres composés gazeux potentiellement toxiques issues de la fermentation des algues vertes (ulves). Mesures réalisées le 13 août 2009 à St Michel en grève. Rapport d'étude. INERIS, France.

Koutsouris, A. (1999) Organisation of extension services in Greece. *Options Méditerranéennes, Série A: Séminaires Méditerranéennes* 38, 47–50

Koutsouris, A. (2008) Learning and Innovation: Still an enigma? Failing to introduce ICM in a Greek village. *International Journal of Environmental, Cultural, Economic and Social Sustainability* 4, 77–84.

LTA and BV-LDG. (2010) Projet de territoire à très basses fuites d'azote-Programme d'actions pour les bassins versants de la Lieue de Grève. Bassins Versants de la Lieue de grève. Available at: http://www.lannion-tregor.com/

Ménesguen, A. (2003) Les 'marées vertes' en Bretagne, la responsabilité du nitrate: IFREMER. Available at: http://archimer.ifremer.fr/doc/00000/143/

Ruhland, M. (2011) Tiefes Tal, tiefer Graben: München kommt sein wasser kostenlos aus dem Mangfalltal, Bürgern und Politikern in der dortigen region missfällt das. In Süddeutsche Zeitung, 11 April 2011. Available at: www.brunnenverlegen-muenchnerwasser.de

SCE (1999) Lutte préventive et curative contre la prolifération des marées vertes. Conception d'un programme d'actions preventive. Description et analyse du réseau socio-économique Saint Brieuc. Conseil Général des Côtes d'Armor.

Vatn, A. (2005) *Institutions and the Environment*. Edward Elgar, Cheltenham, UK.

Vatn, A. (2010) An institutional analysis of payments for environmental services. *Ecological Economics* 69, 1245–1252.

Wilson, G.A. (2007) *Multifunctional Agriculture: A transition theory perspective*. CABI, Wallingford, UK.

Chapter 9

On-farm renewable energy: a 'classic case' of technological transition

L-A. Sutherland[1], S. Peter[2], L. Zagata[3]

[1]*James Hutton Institute, Aberdeen (lee-ann.sutherland@hutton.ac.uk);* [2]*Institute for Rural Development Research, Frankfurt am Main;* [3]*Czech University of Life Sciences, Prague*

Introduction

On-farm renewable energy production can be considered a 'classic case' in relation to socio-technical transition studies, as it involves a socio-technical niche which remained largely unrecognized by both the agricultural and energy regimes until landscape pressures opened up a 'window of opportunity' in the 1990s. Over the 2000s, the niche has become increasingly integrated into both the agricultural and energy regimes. In this chapter, we analyse the transition process of on-farm renewable energy production, through case studies in Germany, the United Kingdom and the Czech Republic, focusing particularly on the role of renewable technologies in transition within the agricultural regime.

The multi-level perspective (MLP) utilized in this research was originally developed to assess technology-based transitions (see Darnhofer, this volume, for a discussion of the MLP). The basic structure of the MLP comes from evolutionary economics and science and technology studies (Geels and Schot, 2010). These foundations are reflected in the emphasis on 'socio-technical systems', which place the interaction of technologies within the context in which they originate and evolve. Within the MLP, major or 'radical' technological innovations are conceptualized as occurring within niches, whereas incremental changes occur at regime level, as the regime responds to pressures from the landscape and other regimes. For radical technological innovations (examples include transition from horses to automobiles, Geels, 2005; and the introduction of nuclear and renewable energy into the electricity production regime, Verbong and Geels, 2007) the focus is on the innovation, evolution and diffusion of technological innovations, typically within a single regime. However, proponents argue that the source of the innovation is usually located either on the periphery or outside of the regime which it eventually influences; owing to the path dependencies of the regime in question, radical innovations occur at the margins, requiring the protection of niche actors in order to develop (Geels and Schot, 2010). In this chapter we address a socio-technical transition which (in Europe) originated in the agricultural sector but which is leading to a radical transition in the energy sector (in other words multi-regime interaction). We assess it as a transition in progress, emphasizing the important influence of policy on the rapid increase in renewable energy production in general, and on farms in particular.

Methods

Production of on-farm renewable energy was studied in three European case study sites, focusing on two specific types of renewable technologies: biogas production through anaerobic digestion in the Czech Republic and Germany; and wind energy production in the United Kingdom (specifically Scotland) (Fig. 9.1). The selection of two different technologies for study reflected the geographical differences between the two regions; the unsuitability of local conditions for maize production make anaerobic digestion much less viable – and therefore less common – in Scotland than in Germany and the Czech Republic. In all three regional study sites, farmers are the most numerous producers of renewable energy using these technologies, although wind energy in the UK as a whole is predominantly produced on large corporate and offshore wind farms.

Both wind and anaerobic digestion technologies produce electricity, although anaerobic digestion also produces heat, bio-methane and processed manure, which can be used as fertilizer. Anaerobic digestion produces biogas through a process of fermenting manure or other residual materials, supplemented by selected crops (typically maize). Raw biogas can be utilized in combined heat and power engines (CHP), or refined into bio-methane fed into the natural gas grid as well as delivered to biogas fuel stations for vehicles. Both technologies have also experienced considerable price supports in the 2000s, leading to exponential increases in uptake. Analysis of these complex processes can thus inform understanding of transition processes.

Data was collected in each country by reviewing regional and national agriculture and energy policy documents, and through interviews with representatives of regional and national governments, the farming industry, landowner groups, lending institutions, regional chambers of commerce, energy consultancies, national parks and young farmers. The focus of the research was on understanding how and why these technologies have become increasingly used on farms in recent years, and the implications for the agricultural industry.

The study sites for this research were the Vysočina region in the Czech Republic, the Wendland-Elbetal region in Germany and the Aberdeenshire region in Scotland. These regions cover areas of 6,795 km^2, 2,020 km^2 and 6,313 km^2 respectively (Fig. 9.1). Agriculture is the primary land use in all three regions, and a major component of their regional economies. Farms in all three regions have mixed structures. In the Wendland-Elbetal region, these are polarized between small-scale farming to the west of the Elbe river and large-scale (500 to 2,000 ha) to the east, a legacy of the former German Democratic Republic's collectivized system. The average farm size in the region is 100 ha, with 60% having less than 75 ha (Bioenergieregion Wendland-Elbetal, 2012). In Aberdeenshire, land is held through a mix of owner occupancy, formal tenancies (long-term rental agreements), contract farming and short-term rental. The majority of agricultural land is held in holdings of more than 200 ha, leading to an estimated average (active) farm business size of 180 ha (Aberdeenshire Council, 2009). In the Vysočina region, the majority of land owners are small-scale (46% are less than 5 ha), with 23% in the 10 to 49.9 ha category. There are over 400 'large' land owners, with more than 100 ha. Most of the land on these farms is rented, owing to land redistribution in the 1990s. In all three sites, renewable energy is most likely to be produced on larger farms, although generation can occur across a range of farm sizes.

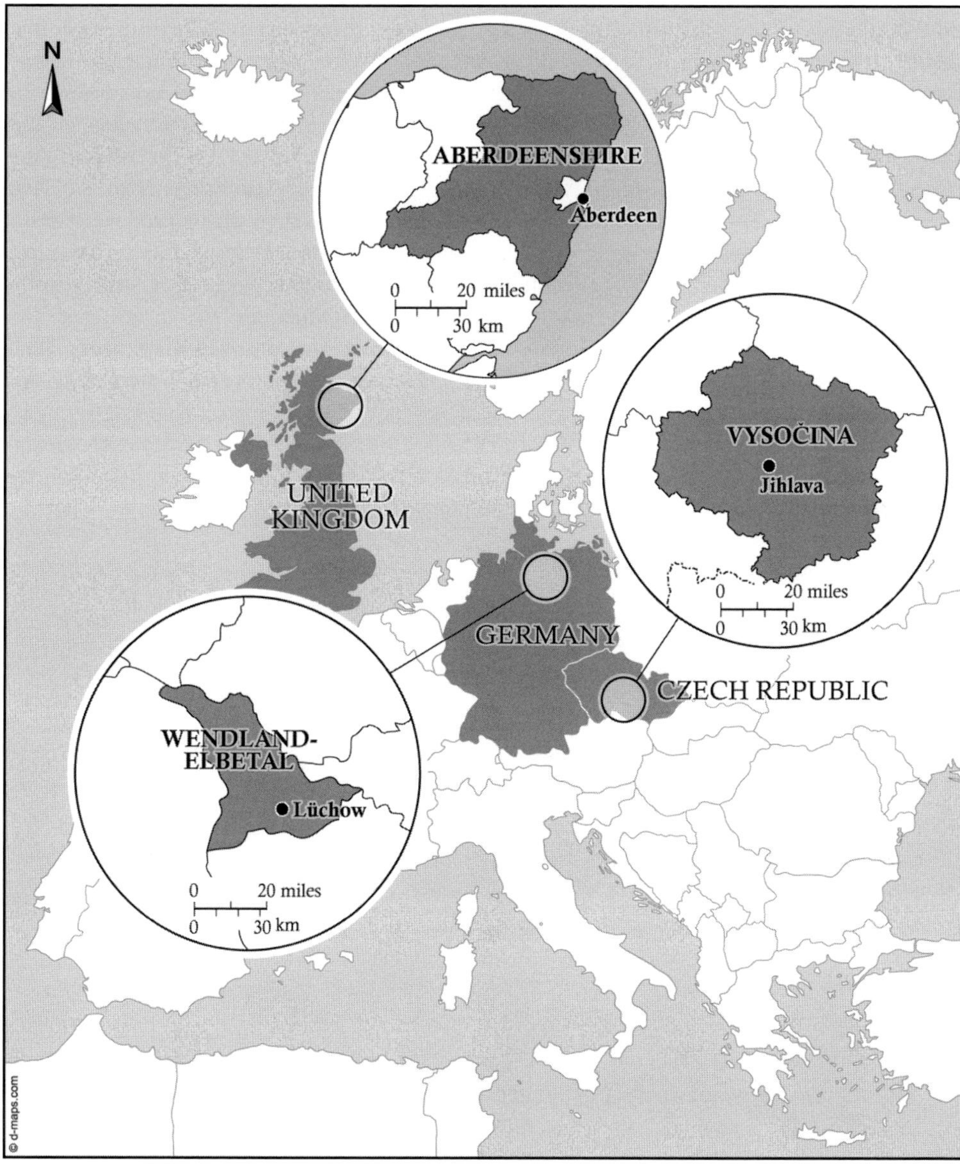

Fig. 9.1. Location of the three case studies: Vysočina, Wendland-Elbetal and Aberdeenshire (Source: authors).

Conceptualizing renewable energy transitions

In relation to the conceptual underpinnings of the research, presented in Chapter 2 (Darnhofer, this volume), on-farm renewable energy production is conceptualized here as representing a niche within the renewable energy sub-regime. This sub-regime was created by the overlap of the agricultural and energy regimes. From the multi-level perspective (MLP), regimes are conceptualized as dominant sets of rules, which coordinate and guide activities in particular directions. Although regimes do not necessarily correspond to sectors, in this case we have identified the agricultural regime (the rules, artefacts and practices characterizing mainstream agricultural production) and the energy regime (the rules, artefacts and practices characterizing mainstream energy production) as the two regimes involved. For biogas, which originated from processing waste products which included non-farm waste, a waste management regime could be included. On-farm renewable energy production is conceptualized as a sub-regime because it satisfies the function associated with the energy regime (energy production) while utilizing the resources (particularly land) and involving actors from the agricultural regime. The technologies and actors associated with renewable energy production on farms are conceptualized as a niche (Fig. 9.2).

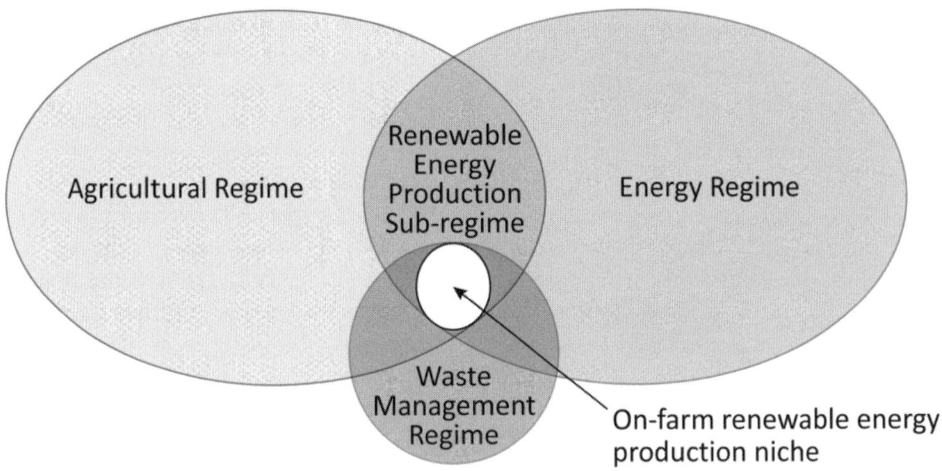

Fig. 9.2. On-farm renewable energy production within the regime structure (Source: authors).

There is a similar structure to the energy regime in all three study countries, whereby most of the energy is produced through large power stations (gas, oil, coal and nuclear), distributed through a centralized international grid system, and regulated at national level. There is also some limited local distribution and consumption. Electricity produced through wind and anaerobic digestion is distributed through this grid system.

Although the agricultural regime is marked by increasing scale and intensification of the production of agricultural commodities, in comparison to the energy regime, production is highly dispersed, with commodity producers numbering in the thousands rather than the tens or hundreds characteristic of the energy regime. Agricultural distribution channels are also more decentralized and complex, owing to the larger number of types of commodities produced, although most of these are vertically integrated (sold through centralized channels), much like the energy regime.

The transition pathway that renewable energy production has followed can be divided into a number of overlapping phases (pioneering, anchoring, translation, take-off, and contested transition) which are discussed in more detail in the following sections.

Pioneering phase (1950s to 1970s)

In the MLP, innovations are conceptualized as occurring as technological novelties; when these novelties gain a supportive or protective environment (for example actors that encourage their development), they are considered niches (see Darnhofer, this volume). These novelties and niches often exist for years if not decades before a 'window of opportunity' opens at landscape level, which leads to increased demand for the technology and thus direct interaction with the regime (Geels and Schot, 2010). We therefore term the development of the niche in its early stages 'pioneering'.

Both wind energy and anaerobic digestion have lengthy histories in terms of development in the study sites. In Germany, anaerobic fermentation was first used after World War II, with approximately 50 to 70 plants in Eastern and Western Germany. Due to oil and coal supply improving during the 1950s, biogas technologies were abandoned and then taken up again in the 1970s, when the oil price crisis led to a revival of interest. In the Czech Republic, anaerobic digestion was first experimented with on farms in the 1970s but the focus was on dealing with the waste associated with large-scale intensive livestock production and heat was simply treated as the by-product of a waste-management solution. The first biogas station was constructed in Czechoslovakia in 1974 and still functions to this day. Wind turbines were also experimented with in the United Kingdom in the 1970s, as part of a search for alternative energy sources in response to the oil crisis of 1973. Prior to that point, wind turbines had been located primarily on farms and rural residences as a means of pumping water, grinding grain into flour and providing electricity to remote farm locations (Gipe, 1995).

In the pioneering 'niche development' phase, it is notable that: (i) the technologies were developed originally for farm use; (ii) the technologies were used to address issues other than climate change (e.g. waste management and to provide alternative energy sources); and (iii) the changes were inspired in part by rising oil prices (a landscape pressure). As such, renewable energy in the 1970s represents what Geels and Schot (2010) would consider a 'transformation pathway', whereby a moderate landscape pressure (peak oil prices) occurs at a time when the niche-innovations have not yet been sufficiently developed to capitalize on this opportunity. Energy regime actors responded by modifying their research and innovation activities (by investing in research into renewable energy production). Subsequent reductions in oil prices and technical difficulties slowed uptake but a foothold had been established. This process of 'anchoring' is further described in the next section.

Anchoring phase (1970s to mid-1990s)

The oil crisis of 1973 led to processes of 'anchoring' of wind and biogas technologies in the case study areas. Elzen *et al.* (2012:3) define anchoring as:

> the process in which a novelty becomes newly connected, connected in a new way, or connected more firmly to a niche or a regime. The further the process of anchoring progresses, meaning that more new connections supporting the novelty develop, the larger the chances are that anchoring will eventually develop into durable links.

These durable links represent long-term connections that eventually lead to regime change. Key in this description is the iterative process of change – the novelty, niche and regime are all changed through their interactions. Elzen *et al.* (2012) describe three types of anchoring: institutional, technological and network (see also Darnhofer, this volume). The establishment of research and development support for the technological development of biogas and wind energy production represents economic institutional anchoring: new rules of practice associated with economic institutions. This process was evidenced in the 1980s, when agricultural science institutions in Germany started putting a stronger focus on working with biogas technologies (Umweltbundesamt, 2010), 'professionalizing' the experiments that had been undertaken at farm level. It was at this point that the discovery was made that adding organic matter can increase energy output from digestion. The UK, German (and Danish and Swedish) governments also invested in wind technology research and development in the 1970s and 1980s (Gipe, 1995). These latter steps can be interpreted as evidence of technological anchoring: further development of the technologies to enable commercial electricity production.

The early 1990s saw a shift in policy support from research into market creation for renewable energy production in the UK and Germany. In the UK, the 1990 Non-Fossil Fuel Obligation required public electricity suppliers to purchase electricity from non-fossil fuel sources which included both nuclear and renewable energy. However, this policy had limited success. In contrast, the Electricity Feed-in Law introduced in Germany created price supports for renewable energy production. This has led to what Bruns *et al.* (2009) define as the '1st departure phase'[1] of the German biogas development, with ongoing professionalization and expansion of the spectrum of actors beyond pioneering farmers and sectoral businesses, to include the waste industry, industrial biogas plant operators and energy suppliers.

This period also saw the first signs of institutional anchoring within the agricultural regime: state support for the development and installation of renewable energy technologies on farms. Although the money involved was considerably smaller than for research and development in the energy sector, it is important to recognize that the first farm diversification grants (which included installation of renewable energy) were put in place by the German and UK governments in the 1980s. These efforts appear largely separate from actions in the energy sector, suggesting that the two regimes were acting independently. Whereas the aim of farm diversification grants was to increase farm household income, research and development activities in the energy sector specifically targeted the development of renewable energy technologies. Owing to a lack of clear energy policy from the Czech state, limited investment was made in Czechoslovakia in

[1] The original German terms used by Bruns *et al.* (2009) were translated into English by the authors of this chapter.

either agriculture or renewable energy production sectors during this period. Anaerobic digestion technology continued to be used primarily for waste-management.

Whereas anchoring was occurring at national levels in Germany and the UK, the Czech Republic thus remained in a pioneering phase, with one of the first digesters installed on a farm in the mid-1990s. The installed technology failed, threatening the local environment, and received heavy public criticism (Klastr Bioplyn, 2013). Two of the regional study sites were also entering the pioneering phase, with individual turbines established on a few farms in Aberdeenshire in the 1980s, and experimentation with biogas in Wendland-Elbetal. While the first digester set up in 1995 was successful in Wendland-Elbetal, in the Aberdeenshire case, the turbines were abandoned in response to technical failures.

Translation phase (late 1990s)

Political action taken to address climate change in the late 1990s represents a 'translation' of the original purposes of biogas and wind energy technologies, from waste management and energy production solutions, to a means of addressing climate change[2]. The European Commission published a White Paper in 1997 that set the first target for renewable energy source (RES) utilization in the EU; the contribution of RES to the gross inland consumption of (primary) energy was to increase from 5.4% in 1997 to 12% in 2010 (European Commission, 1997). The subsequent directives (Renewable Electricity Directive 2001/77/EC and the Biofuels Directive 2003/30/EC) set indicative targets for the RES share in the electricity and transport sectors in 2010. The RES share in electricity consumption should increase to at least 21% in the EU-27, broken down into targets for individual Member States (Kleßmann *et al.*, 2011). In response to the new directives, Germany adopted a renewable energy law in 2000 that secured feed-in tariffs for 20 years for renewable energy and obligated grid operators to connect to renewable energy plants (*Erneuerbare-Energien-Gesetz* – EEG). In 2000, the UK Government announced a target of 10% of electricity to be produced from renewable sources by 2010. In contrast, the Czech Government continued its support for conventional energy sources, with a clear priority of finishing construction of the nuclear powerplant JET, which started to operate in 2002.

As such, the European directives created a 'window of opportunity' for renewable energy development in the UK, and as will be seen in the next section, latterly in the Czech Republic. However, Germany had already accelerated production through a series of national and regional measures. As early as 1997, the regional government in the Wendland-Elbetal region set the goal of meeting 100% of primary regional energy demand locally through regionally produced renewable energy and energy savings. There is no evidence of regional activity relating to renewable energy in the Vysočina or Aberdeenshire case study areas during this period.

Take-off phase (early 2000s)

In all three countries, policy measures undertaken at national level to respond to EC level directives had a considerable impact on the development of renewable energy production,

[2] The term translation is utilized in actor network theory to refer to a process by which the objectives of one actor are transferred onto others, thus recruiting them to the primary actor's network (Callon *et al.*, 1986 in Elzen *et al.*, 2012). Translation is defined differently in other applications of the MLP.

both on- and off-farm. Bruns *et al.* (2009) define the period from 2000 to mid-2004 as a 'boosted departure phase' in Germany. The replacement of the Electricity Feed-in Law by the EEG in 2000, alongside an increase in feed-in compensation, accelerated development in the biogas sector. In 2001 the introduction of the Biomass Regulation (*Biomasseverordnung*) led to the EEG now primarily serving energy and climate objectives instead of waste industry interests (Umweltbundesamt, 2010). The period from mid-2004 to end-2006 marks a 'boom/take-off phase': in 2004 the 1st EEG amendment led to an increase in the number of biogas plants as the renewable materials ('*Nawaro*') bonus made energy crop cultivation (especially maize) highly attractive to farmers. Low agricultural prices also led to the expansion of energy crop cultivation. During the period, professionalization of the biogas sector progressed further. Graded rewards according to output favoured biogas plants of up to 500 kW, but due to unclear definitions it was possible to set up large facilities in the form of 'biogas parks' (consisting of multiple 500 kW plants).

In Scotland, production of electricity from wind energy increased from 216 GWh in 2000 to 2,023 GWh in 2006 (Scottish Government, 2013). This reflects the introduction of renewable obligation certificates (ROCs) in 2002, which required energy companies to produce either a set percentage of their energy through renewable sources, or purchase certificates of renewable energy production from others. This mechanism made wind energy production economically viable at both corporate and large farm scales.

Support for renewable energy production was put in place somewhat later in the Czech Republic than in Germany and the UK, and was heavily influenced by entrance into the European Union. During the pre-accession period (2000 to 2004), the Czech Republic agreed an 8% target for renewable energy production by 2010. In the Czech Republic a discernable 'wave' of biogas production occurred from 2000 to 2005, when German and Austrian companies saw the economic potential to develop anaerobic digestion in response to the Czech state's renewable energy commitment. In 2002, the Czech Government started supporting renewable energy production through price supports, but the stimulation effect remained small (Vitaskova, 2012).

At regional level, institutional anchoring was well underway in the Wendland-Elbetal region, with the awarding of a Sustainable Energy Europe award to the district in 2000, a detailed regional energy use study, and participation in the German 'Regional Action Programme'. A detailed study of regional energy use and potential as well as the elaboration of a conversion scenario were undertaken in 2001. In Aberdeenshire, no specific regional policies were developed but the first commercial on-farm wind turbines received planning permission in 2004, granted to a farmer who had unsuccessfully pioneered the technology in the 1980s. In the early 2000s, the Czech region of Vysočina launched a new grant scheme focused on support for increasing usage of renewable energy sources in the region. Such financial support directly resulted in construction of six agricultural biogas digesters. This grant scheme, therefore, represented one of the first institutional anchorages of alternative energy production at regional level in the Czech case.

Contested transition phase (mid-2000s to present)

As the technologies have proliferated and been integrated into the energy and agricultural regimes, new landscape pressures and regime challenges have arisen. The significant increase in cereal prices in 2007/2008 caused a deceleration of uptake in the biogas sector in both Germany and the Czech Republic. Public criticism regarding bio-energy spread through the media (the food versus fuel debate), and biogas was increasingly questioned in terms of its efficiency and CO_2 balance. This downturn was counterbalanced in 2009 by a new EU Directive on the promotion of the use of energy from renewable sources (RES) (2009/28/EC), which amended and repealed the two previous Directives (European Commission, 2009), and led to a resurgence in new biogas plants in Germany and the Czech Republic. The new Directive set binding national targets for all Member States to reach an overall RES contribution of 20% in the EU's gross final energy consumption, by 2020 (Schwarz *et al.*, 2012). Binding national targets were set at 13% in the United Kingdom (20% in Scotland), 18% in Germany and 13.5% in the Czech Republic.

In the Czech Republic, a second wave of biogas production (2009 to 2013) occurred in response to this increased state renewable energy target but within a more established (stricter) regulatory environment than previously. The period has also seen Czech companies lobbying the government for support and attempting to gain market share through production of associated technologies. Biogas became the fastest growing technology in the Czech RES field, contributing 15.9% of the overall RES production in 2012 (CZBA, 2013). Included in the new regulations are environmental impact issues, including uses for the heat generated through anaerobic digestion (in the first wave, heat was a seldom-considered by-product). This has led farmers to consider business diversification options to make use of the heat. However, the boom for RES technology collided with the economic potential of the country to financially support the transition. Uncontrolled growth of photovoltaics resulted in opposition of the RES in general, on the basis that they increased electricity prices. Although a new Act on RES was passed in 2012, setting up new and innovative ways of integrating biogas technology on farms (including organic farms) and utilization of waste heat (for example for drying crops, heating greenhouses), the Czech Government decided in 2013 to discontinue support for RES once their 2020 target is met. Renewables in this study site represented 10.6% of energy consumption in 2011, an increase from 2% in the 1990s.

In Germany, the amendment of the EEG in 2009 led to another acceleration in the development of the biogas sector. Changes in the compensation system favoured a more effective use of manure and residual materials, and bonus payments for smaller plants with up to 500 kW were introduced (and funding for large 'biogas parks' was eliminated) (Bruns *et al.*, 2009). Within the framework of the 'Energy Turnaround' (*Energiewende*), the 'Energy Concept 2050' (*Energiekonzept 2050*) was enacted in 2010 with the overall objective of achieving an energy supply primarily from renewable sources by 2050. The Fukushima nuclear disaster led to further resolutions in 2011. The EEG's 3rd amendment in 2012 supported facilitation of 'mini' biogas plants of up to 75 kW, which encouraged the establishment of farmer-operated plants for the purpose of using on-farm biomass. In 2011, there were ca. 7,000 biogas plants throughout Germany, and the share of renewables in primary energy consumption rose from 1.3% in 1990 to 10.9% in 2011 (BMU, 2013).

In Scotland, Feed-in Tariffs (FiTs) were introduced in 2010, specifically targeting developments of less than 5 MW. The measure is thus oriented towards smaller-scale developments, including those sponsored by communities and farmers. The Scottish Government also established a Renewables Action Plan in 2009, which set targets for renewable energy production, specifically including community production. Community collaboration was also encouraged by a 2009 support scheme (CARES) but has had limited up-take to date. Although there has been a rapid proliferation of turbine development companies, the number of manufacturers producing turbines has consolidated and turbines are primarily imported from Europe. Total wind energy produced in Scotland increased from 2,023 GW in 2006 to 8,328 GW[3] in 2012 (Scottish Government, 2013).

At regional level, technological anchoring can be observed in 2006 when Germany's first biogas (fuel) station (for vehicles) opened in Wendland-Elbetal, initiated by a regional farmer. From 2009 onward, network anchoring was further advanced by regional energy management agencies that offered advisory services to businesses, municipalities and private households, and thus contributed to the spread of renewable energy use. In 2009, biogas plants accounted for 45% of the total electricity from regional renewable energy sources and 113% of regional electricity consumption was generated from renewable sources. Network anchoring is also evident in Aberdeenshire, where there has been a clear increase in the number of interactions between farmers (informal visits and information days) and events run by farming organizations. Some of the original pioneers became hybrid actors, running consultancies to assist other farmers (Aberdeenshire), or sitting on the council for the Bio-energy Region Programme (Wendland-Elbetal). In Aberdeenshire, the introduction of FiTs led to an exponential increase in applications for planning permission to put up turbines in 2010/2011 but a higher percentage of large turbine applications were rejected, owing to public concerns about regional saturation (visual impact on the landscape).

Although the three countries thus appear to be on somewhat divergent trajectories, there is a common shift towards community development concerns and capitalization on renewable energy production to meet other societal objectives, such as energy tourism (Germany), community development (Scotland) and local heating (Czech Republic). Saturation effects are evident in the lack of grid capacity in all three countries and increasingly expressed through negative public opinion. This is exacerbated by public perception, in all three countries, that rising electricity prices reflect government subsidies and are, therefore, benefiting wealthier society members. Cognitive or interpretative institutional anchorage thus does not appear to have kept pace with normative institutional anchoring, as is evident in the public protest that has been organized in the last few years. As a result, all three governments are encouraging more collaboration between farmers and local communities but with limited success to date.

In terms of technological anchoring, experimentation continues on biogas technologies (for example on additional substrates and use of waste heat). In the Czech Republic and Germany, farmers continue to experiment with, and develop, anaerobic digestion technology. Agricultural technology suppliers are also becoming involved in biogas technology supply. Farmers are on the executive board of the Regional Action association supporting the Wendland-Elbetal Bio-energy region. The dependence of biogas on agricultural products and by-products has enabled farmers and agricultural regime actors to retain a level of control over the sub-regime and, as a result, these actors are actively resisting penetration by electricity companies. This is not the case with wind energy, where the increasing risks of not acquiring grid access or planning permission are more easily accepted by corporations than by farm businesses (Aberdeenshire is unusual in Scotland for

[3] Including a small amount of wave, tidal and solar energy.

the high number of farmer-owned developments). For wind energy, developments addressing noise reduction and cooling systems are undertaken by manufacturers.

Implications of transition towards renewable energy production

Utility of the multi-level perspective

In our introduction, we suggested that on-farm renewable energy production represents a 'classic' socio-technical transition in that the transition process has followed many of the tenets set out in the MLP. In our case studies, there is a clear progression from niche to anchoring, through processes which redefined the initial purpose of the technology and adapted it to meet an emerging landscape-level need. Where the transition process differs from the MLP as presented in Chapter 2 (Darnhofer, this volume), is that the process does not appear to have resulted primarily from niche actors pursuing a change in the agricultural regime. Indeed, it appears that landscape factors led energy regime actors to pursue renewable technologies and that the agricultural regime was impacted upon because of the resources which were historically within its purview: land but also agricultural products and waste.

There is a clear process of alignment between the energy and agricultural regimes in terms of functions (i.e. energy production) and in terms of the socio-technical system (renewable energy produced by farmers shares the same grid access, planning procedures etc. as that produced by other businesses). However, where this appears to reflect an emergent regime change (based around technological substitution) within the energy sector, the transition within the agricultural sector is relatively minor (but may lead to a transformation pathway). As a result, commodity production remains the primary function of the agricultural regime and energy production is instead viewed as farm diversification. Renewable energy production is thus a case of multi-regime interaction. In their conceptualization of multi-regime interaction, Raven and Verbong (2007) propose multiple types of multi-regime interaction: competition, symbiosis, integration and spillover. In this present research, the energy and agricultural regimes compete for use of agricultural land, both functionally (food vs fuel) and in terms of control (farmer vs corporations). The reliance of biogas technologies on both land and agricultural products has enabled the agricultural sector to retain a stronger foothold than has been the case with wind turbines but there is a spillover effect. The multiple functions of the agricultural sector suggest that it may represent a special case in relation to the MLP, where regimes are defined in terms of a single function (this will be discussed further in relation to all cases in Darnhofer *et al.* and Sutherland *et al.*, both this volume).

The analysis of the three cases also demonstrates the different ways that MLP concepts were enacted. Germany was the first of the three countries to recognize the economic development opportunity represented by renewable energy production, and so was already actively establishing targets – and pursuing foreign developments – before the UK and Czech Republic. The 'window of opportunity' represented by the EU directives, thus, may not have been identified as such if Germany were analysed on its own. The discussion above has also emphasized that change occurs differently at regional level, where pioneering is occurring while anchoring is underway nationally, while niche innovations are frequently located outside of the region under consideration. How these innovations diffuse and how the MLP can be utilized to understand regional level transitions will be addressed in subsequent publications. For further detail on the UK case, see Sutherland and Holstead (2014).

Role of policies

EU, national and regional-level policies have clearly had a major impact on renewable energy development, acting as 'triggers' to increase uptake of the respective technologies in all three case study regions. From the 1970s onwards, institutional anchoring through government policies was evident in the UK and Germany. In Czechoslovakia, national energy policy followed no clear renewables strategy, and to date there has been no relevant policy influence at the level of the Czech study region. 'Internalization' of landscape pressures (such as peak oil and high agricultural commodity prices) is evident in the policy framework supporting renewable energy production within the energy regime. The decisive role of EU, national and regional policies can be seen in the 'boom' during the take-off phase in all three case studies.

While there is a clear climate change emphasis within the renewable energy mandate, it is also apparent that the Scottish and German governments view renewable energy production as an economic development opportunity, whereas the Czech Government's renewable energy policy is anchored in EC directives. Successful regime change may reflect the business opportunity represented by renewable energy production, enabling commercial companies to engage in, and pursue, the transition. Accordingly, market price policy for renewable energy has had a critical impact. Further significant influencing factors are the availability of, and access to, the grid; (support for the) capacities of farmers to invest in the technology; bonus payments for certain substrates in the case of biogas; as well as targeting support at specific scales of facilities (in terms of energy production capacity). In the 'contested transition' phase (mid-2000s onwards), the immediate impact of withdrawal or changes to support underline this policy dependence. The growing debate about sustainability of such supports in all three countries suggests that a suitable public support model has not yet been found, despite the priority of renewable energy production in official policies.

In addressing climate change, governments have found it easiest to follow the traditional 'technological fix' approach, funding renewable energy production. This can be regarded as a form of 'transition management' in the sense of governments seeking to actively encourage renewable energy production but seeming less concerned with the protagonists (see also Wilson, 2012). Only latterly, opportunities for community development have been recognized; that is, there is growing recognition of the relevance of the social dimension for innovation and uptake.

Structural implications

The transition to renewable energy production appears to be a relatively minor transition in the agricultural sector but could nevertheless lead to a 'reconfiguration pathway' in the agricultural regime, defined by Geels and Schot (2010) as symbiotic innovations which subsequently trigger further adjustments in the basic architecture of the regime. In all three sites, renewable energy production most commonly occurs on large-scale farms. This is because these farm businesses have the resources to fund, or leverage, the significant capital investment required for profitable renewable energy generation and in the case of biogas, the capacity to produce sufficient inputs. The decentralization of energy production within the energy regime, as a result of dispersed renewable energy production, could thus contribute to the increasing concentration of production in the agricultural regime. This has sustainability implications, especially as the reliance of biogas technologies on maize suggests that continued up-take will lead to further intensification and mono-cropping in agriculture, although efforts are being made to diversify the crops which can be used in digesters. There is also some evidence in the three sites that organic farmers are taking up renewable energy production, as it is consistent with their low impact orientation. There is also some suggestion that having a viable alternative source of income will mean that farmers feel less economic incentive to intensify their commodity production to make a profit.

Undoubtedly, introduction of renewable energy production on farms gives farmers access to the energy sector, altering their economic position. Farms producing renewable energy can 'process' their produce independently and are thus protected from volatile market prices for agricultural commodities. Alternatively, while low prices have encouraged uptake of biogas, rising commodity prices are providing a disincentive. Energy production on farms therefore offers a potential escape from the squeeze-on-agriculture effect that has been firmly associated with modern productivist systems (van der Ploeg, 2006; Wilson, 2007). However, profitability of the farms producing renewable energy is now vulnerable to the institutional arrangements of the energy regime and its sectoral policies.

Conclusion

In this chapter we have presented three case studies of transition towards on-farm renewable energy production. As 'transitions in the making' it is not yet clear yet what form regime change will ultimately take in the three countries, or indeed if the transition will be 'completed', owing to saturation effects and subsidy dependence. At present, considerable branching appears underway. Analysis using the MLP has proved useful for describing the trajectory to date, with evidence of niche-, regime- and landscape-level factors. However, operationalizing MLP terms for use in analyzing an emerging transition proved challenging (also explored in detail in Karanikolas *et al.*, this volume). The analysis demonstrates the need for further research on multi-regime interaction and spatial location of transitions. The authors also question the utility of defining regimes in response to a single function, in light of the multiple functions characterizing the agricultural sector.

References

Aberdeenshire Council (2009) Agriculture statistics. Available from http://www.aberdeenshire.gov.uk/statistics/economic/

Bioenergieregion Wendland-Elbetal (2012) Die Energiewende in die Köpfe: Kompetenzregion Bioenergie Wendland-Elbetal, Regionalentwicklungskonzept 2012-2015, veröffentlichte Gesamtfassung. GLC Glücksburg Consulting AG, Lüchow, Germany.

BMU (Federal Ministry for the Environment, Nature Conservation and Nuclear Safety) (2013) Zeitreihen zur entwicklung der erneuerbaren energien in Deutschland (Timelines on the development of renewable energies in Germany). BMU, Berlin. Available at: http://www.erneuerbare-energien.de/fileadmin/DatenEE/BilderStartseite/Bilder_Datenservice/PDFsXLS/

Bruns, E., Ohlhorst, D., Wenzel, B. and Köppel, J. (2009) Erneuerbare energien in Deutschland – eine biographie des innovationsgeschehens (Renewable energies in Germany – a biography of the innovation process). Universitätsverlag der TU Berlin. Available at: http://www.opus4.kobv.de

CZBA (2013) Mapa bioplynových stanic (Map of the biogas digesters). Česká bioplynová asociace. Available at: http://www.czba.cz/mapa-bioplynovych-stanic/

Darnhofer, I. (2015) Socio-technical transitions in farming. Key concepts. In: Sutherland, L-A., Darnhofer, I., Wilson, G.A. and Zagata, L. (eds) *Transition Pathways towards Sustainability in Agriculture: Case Studies from Europe.* CABI, Wallingford, UK, pp. 17–32.

Darnhofer, I., Sutherland, L-A. and Pinto-Correia, T. (2015) Conceptual insights derived from case studies on 'emerging transitions' in farming. In: Sutherland, L-A., Darnhofer, I., Wilson, G.A. and Zagata, L. (eds) *Transition Pathways towards Sustainability in Agriculture: Case Studies from Europe.* CABI, Wallingford, UK, pp. 189–204.

Elzen, B., van Mierlo, B. and Leeuwis, C. (2012) Anchoring of innovations: Assessing Dutch efforts to harvest energy from glasshouses. *Environmental Innovations and Societal Transitions* 5, 1–18.

European Commission (1997) *Energy for the Future: Renewable Sources of Energy. White Paper for a Community Strategy and Action Plan.* COM(97)599 final (26/11/1997). Brussels, Belgium.

European Commission (2009) Directive 2009/28/EC of the European Parliament and of the Council of 23 April 2009 on the promotion of the use of energy from renewable sources and amending and subsequently repealing Directives 2001/77/EC and 2003/30/EC. *Official Journal of the European Union*, 5.6.2009.

Geels, F.W. (2005) The dynamics of transitions in socio-technical systems: A multi-level analysis of the transition from horse-drawn carriages to automobiles (1860-1930). *Technology Analysis & Strategic Management* 17, 445–476.

Geels, F.W. and Schot, J. (2010) The dynamics of transitions. A socio-technical perspective. In: Grin, J., Rotmans, J. and Schot, J. *Transitions to Sustainable Development.* Routledge, New York, pp. 11–101.

Gipe, P. (1995) *Wind Energy Comes of Age.* John Wiley and Sons, Chichester, UK.

Karanikolas, P., Vlahos, G. and Sutherland, L-A. (2015) Utilising the multi-level perspective in empirical field research: Methodological considerations. In: Sutherland, L-A., Darnhofer, I., Wilson, G.A. and Zagata, L. (eds) *Transition Pathways towards Sustainability in Agriculture: Case Studies from Europe.* CABI, Wallingford, UK, pp. 51–66.

Klastr Bioplyn (2013) Historie bioplynek (History of the biogas digesters). Available at: http://bioplynrozvijivenkov.cz/klastr-bioplyn/

Kleßmann, C. (2011) Increasing the effectiveness and efficiency of renewable energy support policies in the European Union. PhD Thesis, University of Utrecht, Utrecht, the Netherlands.

Raven, R. and Verbong, G. (2007) Multi-regime interactions in the Dutch energy sector: The case of combined heat and power technologies in the Netherlands 1970–2000. *Technology Analysis & Strategic Management* 19, 491–507.

Scottish Government (2013) Energy statistics summary. Available at: http://www.scotland.gov.uk/Topics/Statistics/Browse/Business/Energy/energysumjun2013

Schwarz, G., Noe, E. and Saggau, V. (2012) Comparison of bioenergy policies in Denmark and Germany. In: Almås, R. and Campbell, H. (eds) *Rethinking Agricultural Policy Regimes: Food*

Security, Climate Change and the Future Resilience of Global Agriculture. Research in Rural Sociology and Development. Emerald Publishers, Bingley, UK, pp. 235–262.

Sutherland, L-A. and Holstead, K. (2014) Future-proofing the farm: On-farm wind turbine development in farm business decision-making. *Land Use Policy* 36, 102-112.

Sutherland, L-A., Zagata, L. and Wilson, G.A. (2015) Conclusions. In: Sutherland, L-A., Darnhofer, I., Wilson, G.A. and Zagata, L. (eds) *Transition Pathways towards Sustainability in Agriculture: Case Studies from Europe.* CABI, Wallingford, UK, pp. 205–220.

Umweltbundesamt (2010) Biogaserzeugung in Deutschland (Biogas production in Germany). SPIN Hintergrundpapier, Dessau-Roßlau. Umweltbundesamt. Available at: http://spin-project.eu/downloads/

van der Ploeg, J.D. (2006) Agricultural production in crisis. In: Cloke, P., Marsden, T. and Mooney, P. (eds) *The Handbook of Rural Studies*. Sage Publications Ltd, London, UK, pp. 258–277.

Verbong, G. and Geels, F. (2007) The ongoing energy transition: Lessons from a socio-technical, multi-level analysis of the Dutch electricity system (1960-2004). *Energy Policy* 35, 1025–1037.

Vitaskova, A. (2012) Zastavme podporu obnovitelných zdrojů (Let's stop the support for the renewable energy sources). Newsletter CEP*,* září 2012, 4–5.

Wilson, G.A. (2007) *Multifunctional Agriculture: A Transition Theory Perspective*. CABI, Wallingford, UK.

Wilson, G.A. (2012) *Community Resilience and Environmental Transitions*. Routledge/Earthscan, London, UK.

Chapter 10

'The missing actor': alternative agri-food networks and the resistance of key regime actors

C. Darrot[1], M. Diaz[1], E. Tsakalou[2], L. Zagata[3]

[1]*AGROCAMPUS OUEST, Rennes (catherinedarrot@gmail.com);* [2]*Agricultural University of Athens;* [3]*Czech University of Life Sciences, Prague*

Introduction

This chapter focuses on alliances and resistances within niches at the regional scale, and on the specific processes which allow them to overcome barriers and enable an emerging transition. The case studies presented in this chapter are alternative agri-food networks (AAFNs) in three regions of Europe: short supply chains in Rennes Metropole (France); farmers markets in Pilsen (Czech Republic); and integration of winemaking and conventional tourism in Santorini (Greece) (Fig. 10.1).

AAFNs are a new form of food production, marketing and consumption initiative based on an increased and more personalized link between producers and consumers. These networks occur most often at a local level and share values of economic and social solidarity, environmental conservation and opposition to the logic of the dominant food-system (Balasz, 2009; Darrot and Durand, 2011; Verhaegen, 2012). Thanks to their ability to answer the societal demand of 're-linking' around food (Watts *et al.*, 2005, 2008), and because of their impressively fast development, AAFNs have been increasingly understood as a social trend (Marsden *et al.*, 2000; Renting *et al.*, 2003; Maréchal, 2008). First initiated by localized small networks described as societal and commercial niches, AAFNs were progressively named and recognized during the mid-2000s as a coherent and significant phenomenon (Spaargaren *et al.*, 2012; Kneafsey *et al.*, 2013) based on 're-linking' behaviours.

AAFNs are mainly defined in opposition to conventional production. Alternative forms of marketing are rapidly developing as a reaction to globalizing trends. They critique the globalized market and its consequences, in particular aggressive competition between international retail chains which has led to price conflicts between retailers and suppliers. To maintain profits, suppliers tend to decrease prices, which are then reflected in the quality of products. As a reaction, an attachment to locality is developing through the promotion of local food products (Colman, 2008).

Given that AAFNs are based on the aim of increasing actor autonomy through local resources and networking, from the perspective of the FarmPath project these initiatives

 Transition pathways towards sustainability in agriculture: Case studies from Europe. (eds L-A. Sutherland *et al.*)

can be understood as interesting indicators of emerging territorial transitions in the framework of farming and food production activities. Using the multi-level perspective (Geels, 2010), AAFNs represent a niche with conventional (long) food supply chains representing the regime. A dedicated cluster of AAFN case studies was therefore chosen as part of this research, with case studies situated in France, Greece and the Czech Republic (Fig. 10.1).

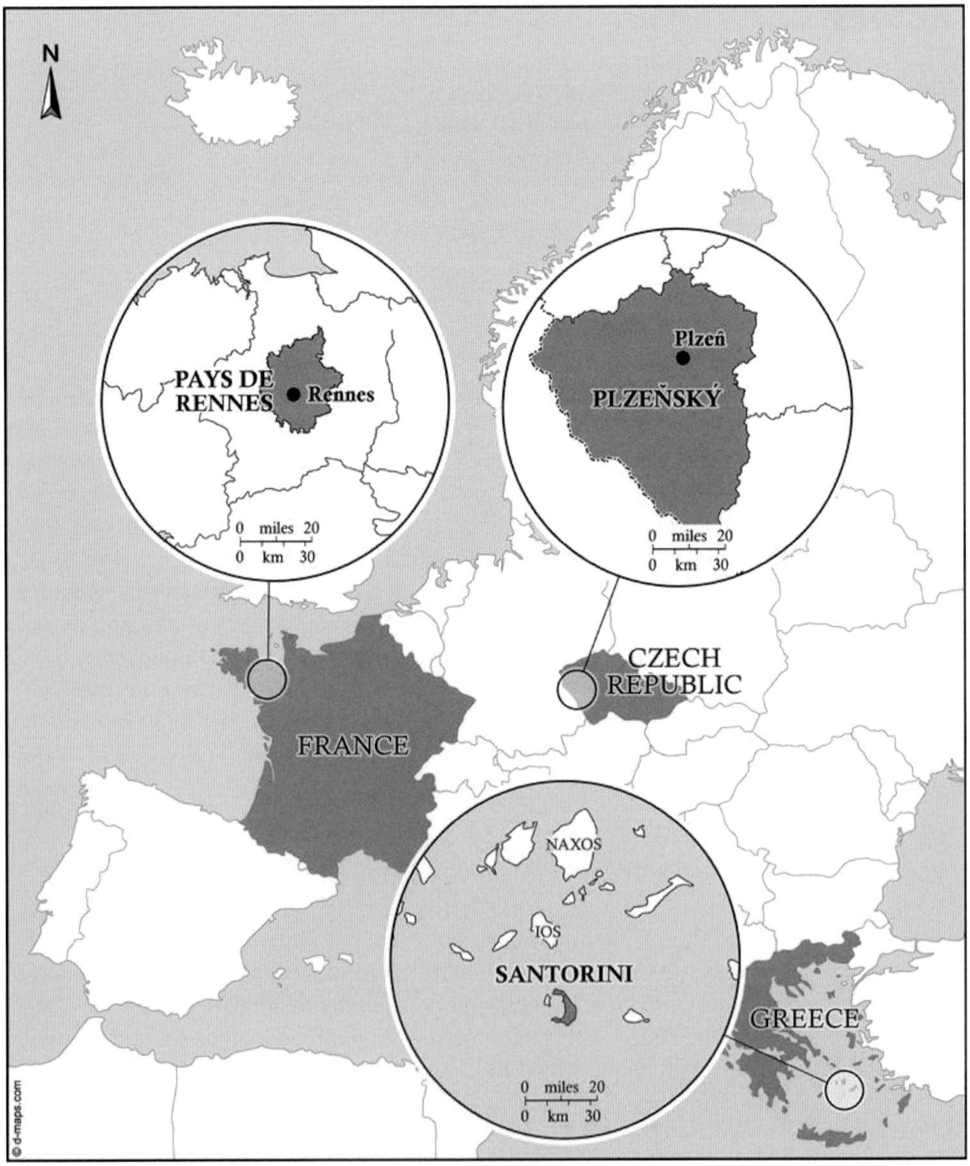

Fig. 10.1. Location of the three case studies: Rennes, Pilsen and Santorini (Source: authors).

The perspective that we develop in this chapter, based on three niches, can be related to the specificity of alternative marketing channels: the framework federating actors which support AAFNs is mainly structured around an alternative vision of economy based on autonomy regarding industrial food production, processing and/or marketing. In other words, the dynamic of those different niches is a multidimensional process of autonomy toward the dominant food regime. In Rennes and Pilsen, the aim is for notions of equity and of direct negotiation in economic relationships between producers and consumers (Darrot and Durand, 2011) to replace the anonymous power within long agri-industrial supply chains. In a similar way, the technical and economic autonomy gained by winemakers in Santorini, thanks to their close partnership with the tourism sector, can be understood as a dynamic cutting the power of international wine market operators, who dominated negotiations around prices and market conditions.

In the three case studies, the notion of quality brings the actors involved in the niche together. Their vision of quality challenges that promoted by agri-industrial chains (based on hygiene rules, standardized products and stable organoleptic food characteristics). On the contrary, the turn to 'quality' (Murdoch *et al.*, 2000) can be understood as both a landscape pressure and a feature characterizing our niches. It relates the notion of quality of social (local product identity, a sense of territory, history, and heritage), ecological (related to environmentally-friendly production or transportation practices), or ethical (social links enabled by the production and marketing process, equity in income and profit share) dimensions. This last element drives the close relationship between the expectations of autonomy and quality, as elaborated by the various types of actors connected around the niche.

We will describe, compare and comment on the emerging transition processes observed during the extension of a wide range of local marketing solutions for farm products in Rennes (France); generalization of farmers' markets in Pilsen (Czech Republic); and local quality wine marketing of Santorini Island wines through the tourism sector (Greece). We will first present the components of these emerging transitions: the sub-systems characterizing the alternative marketing channels, and the characteristics of the niches which initiated the transition. We then analyse the processes characterizing the anchoring and linking of the initiatives in the local dimension of the regime. We pay particular attention to resistance from some of the sub-regimes; how and why they appeared and how they were eventually overcome by networking in the niche. In the concluding section, we reflect on the lessons learned about transition through the analysis of resistance processes.

Overview of the transition

The sub-regimes

Studying AAFNs implies a focus on the agri-food regime. This regime is based on several intertwined functions: agricultural production, processing and marketing (under which we also include consumption habits). These three areas represent the three sub-regimes of the agri-food regime. In Rennes and Pilsen, the dominant regime is characterized by modern industrial agriculture and a processing industry which is directly related to distribution of food via large retail chains (supermarkets and hypermarkets). In Santorini the tourism regime has played an important role in the transition and will be included in our analysis.

The niches and their general trajectories

An interesting, and specific, characteristic of the three initiatives is that each of the niches was initiated within one of the sub-regimes, different in each case but with comparable consequences in terms of the development of the alternative agri-food network. This challenging observation regarding transition processes within connected sub-regimes merits further discussion.

In Rennes, the niche was initiated by farmers within the production sub-regime. In the beginning of the 1980s, the main market in Rennes (which is one of the oldest in France, initiated during the 15th century) was maintained through an alliance between consumers and farmers. The alliance raised a movement against a local authority project to close the large market square and dedicate it to other purposes. A few years later, a group of local farmers opened two farm shops in the suburbs of Rennes. During the 1990s and particularly the early and mid-2000s, an ever larger number of diverse local marketing initiatives developed, such as box schemes, open air markets for local products, farm shops and on-farm marketing. Most of these pioneer initiatives were coordinated by an association of farmers (the 'Regional Federation of the Initiative Centers to Valorize Farming and Rural Areas' (FDCIVAM)) which was dedicated to developing a localized economy based on autonomous farming systems (low inputs, added value retained on the farm) and included ethical values related to environmental protection and social equity. The organization was followed by other producers initiating various forms of AAFNs, and maintaining and reinforcing open air markets. Some indicators therefore show the existence of a long-term transition in the governance of food production. For example, the number of AAFNs registered by the 'Local Observatory of Short Food Supply Chains', managed by the FRDIVAM and the Chamber of Agriculture, continued to increase during the 2000s and by 2011, 77 initiatives were registered. These initiatives share an objective to only market products originating less than 80 km from Rennes and involving only one or no intermediaries between producer and consumer. Based on the limited scale of each initiative, the niche increased its activities and progressively initiated larger and more institutionalized solutions, such as a coordinated system offering organic and local products to restaurants in and around Rennes. This initiative, called 'Manger bio 35' ('eat organic 35' – 35 being the number of the political district) is one of the many AAFNs which can be seen around Rennes. Manger bio 35 was initiated in 2000 by FRDIVAM, in partnership with Rennes City Council and the Regional Council. From an initial group of 14 farmers, the initiative has grown to include 24 farmers.

In Pilsen, farmers' markets are an example of a recent and emerging transition. Farmers' markets are one of the oldest and most widespread forms of direct food marketing (Gale, 1997). However, in the Czech Republic this form of marketing has been significantly challenged since 1989, when the new industrial system of production and consumption was becoming established in Eastern and Central European countries. In 2009, the niche was initiated by consumers in opposition to the marketing sub-regime: the initiators of the farmers' markets developed a critical stance towards food produced on an industrial basis within the 'hygienic-bureaucratic mode' with its specific quality standards

(Marsden, 2006). Some urban consumers took the initiative to organize farmers' markets, mainly by contacting local small-scale subsistence-style producers and convincing them to come and sell their products in the city markets that were created for that purpose. This initiative was very successful, and just 2 years later there were more than 200 farmers' markets in the Czech Republic (including Pilsen). The total turnover reached 1.5 billion CZK in 2011 (1 euro = 25 CZK). This figure is still very low compared to the financial power of the retail chains but it shows the rapid progression of farmers' markets, which initiated important changes in consumer demands. Consumers thus discovered food quality as defined by product and process based qualities. This evolution was accompanied by a shift in the relationship between producers and consumers, linking rural and urban areas. After those first years, farmers can now choose at which of the various farmers' markets in Pilsen to offer their products.

In Santorini, the transition was initiated by winemakers working with the processing sub-regime, and in close partnership with the tourism regime. Historically, wine was made in small-scale peasant winemaking installations. After 1970, with the rapid growth of tourism, farmers preferred to be engaged in tourism and in professions that provided greater and more reliable income. Although grapes were still produced, winemaking activities and the wine trade had generally ceased. Wine was thus no longer made by grape growers, but by private wineries and by the Union of Cooperatives of Santorini ('SantoWines'). During the 1980s, a well-known national winemaker took advantage of the specificities of Santorini wine and started quality improvement experiments. This individual developed a new winery on the island, based on new winemaking techniques. The winery was built in an area with vineyards, remote from tourism but close to the archeological site of Acrotiri. It combined facilities for the production and bottling of quality wine, along with facilities which offered guided tours to tourists. Almost at the same time (in 1990), SantoWines constructed a new winery and began to bottle wine, 2 years later creating a wine tourism centre. These developments also involved new entrants into agriculture who came back to the island to produce quality wine. The purpose of the change was not only to improve the wine, but also – or in fact mainly – to meet tourists' expectations concerning both wine quality (less alcohol and acidity, consistent quality, storage quality, better packaging) and the rising demand for oenotourism (visits to wineries coupled with cultural and historical places and events).

A point of commonality between the three cases is that although the niches were clearly initiated within one of the sub-regimes (producers in Rennes, consumers in Pilsen, processors in Santorini), partnerships with actors representing the sub-regime or even another regime (tourism in Santorini) were absolutely necessary from the start. This lead to 'niche tandems': producers enrolled consumers in Rennes; consumers enrolled producers in Pilsen; and processors enrolled the tourism industry in Santorini.

In Rennes, the construction of the niche was initiated in close partnership with consumers: the producers who initiated alternative and local marketing solutions had managed to meet the expectations of militant urban consumers (the so-called 'consum'actors') who used food choices as a means to express their convictions regarding environmental and economic solidarity.

In Pilsen, collaboration between consumers and local farmers was necessary. The organizers of the first farmers' markets described the initial lack of confidence amongst the farmers, and their struggle to convince enough farmers to run a successful market.

Now, farmers can choose from a range of different farmers' markets to sell their products and negotiation processes appear to have become key elements of the niche. For example, organizers visit farmers and invite them to choose 'their' market. The consumers involved in the niche often share lists of farmers and information about them. The farmers also share information about market organizers and their approaches. The most popular actors in the initiative can freely choose with whom they want to cooperate.

In Santorini, the winery visits and the new type of quality wine directly marketed on the island could not have been initiated and stabilized without close partnerships with actors in the tourism sector. The niche created new networks, especially with the tourist sector, which until that time existed in parallel with agriculture, sometimes depriving agriculture of land and labour. To date, there are two specialist tourism agencies that provide exclusive wine tours of the island. Most wineries have arrangements with tour operators to include their facilities in the tours, as well as individual agreements with hotels to direct guests to wineries as an alternative tourism experience. Without such partnerships, the first steps of niche development as well as its first institution-based anchorage would have failed.

Initial anchoring

In all three case studies, the first steps of anchoring with the regime were initiated by the niche tandems described above, building a strong link between two sub-regimes. Put another way, the niche, understood as a 'locus of radical innovation' (Geels, 2010), was limited during the first steps of anchoring to a few of the sub-regimes involved in transition processes initiated by AAFNs: in Pilsen and Rennes, only two of the three sub-regimes were involved; in Santorini only one of the three sub-regimes was involved, in association with another regime (tourism).

Different processes played a converging role in the first steps of anchoring, and illustrate the three dimensions – technical, network and institutional anchoring – elaborated by Elzen *et al.* (2012). A common perspective on these anchoring processes is provided by the development of an alternative and holistic vision of quality, and of a strategy of increased autonomy toward the agri-food regime. The first steps of anchoring were enabled by the behaviour of niche initiators.

In Pilsen and Santorini, technological and institutional anchoring was enabled by actors bringing new ideas or technical innovations into the food sector: new food marketing solutions in Pilsen and new techniques of winemaking in Santorini. In both cases, the main network anchoring was based on activities undertaken by 'outsiders' (actors who originally did not belong to the sector). However, in Pilsen these actors either had a close relationship with gastronomy (conferences, competitions, exhibitions) or were engaged in experiments with different AAFNs (such as box-schemes and consumer groups). Organizing a farmers' market was, thus, a continuation of previous efforts which reflected an anticipated social demand. In Santorini, the well-known winemaker who initiated the production of quality wine along with offering tourism services was a new entrant in the area but had long experience of winemaking in Greece. The two individual winemakers who followed were new entrants into agriculture, returning to their home region after studying in urban centres. New entrants have also come from traditional family businesses. Together, these stakeholders have created the 'Winemakers

Association of Santorini'. Most of the winemakers, and particularly the pioneers, are newcomers to the area, or new entrants into winemaking. In both Pilsen and Santorini, the innovations were successful enough to rapidly spread at the local level, building on a renewed vision of food quality and on a vision of technical and economic autonomy.

Further in the anchoring process, the networking dimension has played a significant role, particularly in reinforcing the links between the initiators of the niche and their main partners, the tandem of niche actors mobilized in the transition. In Rennes, alternative farmers involved in the niche gathered in several regional-level associations in an official network of associations called 'INPACT pole'. This network played a strong role in the diffusion of new technical references such as organic farming or cattle breeding, based on low external input and reduced cash flow. This network also initiated or progressively promoted a large range of initiatives to re-link producers and consumers at the local level (e.g. farmers' markets, box schemes, internet networks for on-farm marketing dedicated to consumers).

A similar dynamic was observed in Pilsen, where internet networking played a crucial role in the diffusion of references and information on niches: consumers initiating farmers' markets created several online platforms that provided up-to-date information for consumers, including an ever growing list of producers ready to sell either on-farm or at farmers' markets, and information concerning products. The social networking site Facebook also provided important support to the niche, first in networking initiators of the market and also particularly as a vector of rapid political demands. A Facebook site called 'We want farmers' markets in Prague' was aimed at convincing local authorities to support the organization of farmers' markets. This social media dialogue was also set within the context of local elections in 2011 and resulted first in a positive political response during elections and second, in the engagement of the new mayor in supporting farmers' markets at the regional level. An event that was decisive for the reinforcement of anchoring of the niche was the appointment, 1 year after the elections, of the Mayor of Prague as Minister for the Environment. The Ministry of Environment set up a grant scheme in 2011, which has become a key opportunity for initiative development and anchoring in the regime. This scheme is a strong indicator of institutional anchoring and the direct result of networking.

In Santorini, SantoWines also played a decisive role in the diffusion of the technical references supported by the niche. SantoWines, one of – if not the main – promoter of the niche, has expanded and is now the biggest wine producer on the island. Amongst its activities, SantoWines offers advice to farmers and handles subsidies on their behalf.

Overcoming resistance: towards a widespread linkage

Enrolment of new actors

Transitions do not necessarily proceed smoothly. Although resistance from the regime is well documented and understood as part of the transition process (see Darnhofer, this volume), resistance within or around the niche itself is less well understood and described. We will, therefore, focus specifically here on alliances within the niche at the regional scale, and the processes which enable resistance to be overcome and allow the transition to proceed.

Our three cases provide heuristic illustrations of such resistance processes at the niche level. To reinforce the first dimension of anchoring, the actors involved in the tandem which initiated the niche had to manage and enrol a third category of actors. In Rennes, the producer-consumer tandem found it necessary to find intermediary processors willing to be involved in the AAFN. Three examples exemplify this process. First, the Manger Bio 35 group (the network of farmers offering organic produce for public procurement in places such as schools and hospitals in and around Rennes) had to work together to handle the larger amounts of produce needed by these venues, raising the need to create sophisticated logistics to collect products from various farms. In the second example, beef farmers had to find slaughter houses willing to return processed carcasses back to farmers, who then needed to find a butcher willing to prepare meat in packages which were adapted to direct marketing. In the third example, farmers who marketed their products at various local markets and in box schemes began to organize an association and employed an individual to collect and sell the produce on their behalf at the market (at least one producer must be present to comply with legal requirements). Without such solutions, direct marketing of farm produce would not be feasible for a single farmer working alone.

In Pilsen, the size of the niche has remained limited enough to allow food processing either at the farm level before marketing, or by the consumer. Associations to support collective marketing and supply have not yet developed in Pilsen, but the rapid speed of niche development may initiate similar developments to those in Rennes, enrolling intermediary actors. This approach would allow the group to cope with the growing volume of local quality food produced in the territory and avoid the difficulties of time constraints for farmers involved in direct marketing.

In Santorini, the wine-tourism sector tandem needed the commitment of farmers to grow grapes. Farmers were expected to change their production practices to meet new quality standards required by the niche. These changes included: earlier grape harvesting (August rather than September) to lower the degree of alcohol and the acidity of the wine produced; harvesting all red and all white grapes at one time, instead of daily harvesting of mature grapes; growing selected local varieties and only one variety per parcel instead of multiple vine types per parcel; and more recently, vine cultivation in lines instead of scattered plants to facilitate access by machinery.

Further, the third set of actors in this case expected – and even pushed – to join the niche in order to pursue the successful trajectory initiated by the original tandem. This third set of actors had shown active or passive signs of resistance to this enrolment.

Why was there resistance?

In Rennes during the 1980s and early 1990s, and in Pilsen during the early stage of the niche, intermediary actors (representing the processing and logistics sub-regime) were absent. Niche actors did not try to enrol them, but neither did the 'missing' actors try to join the niche. Their absence can be considered a strong expression of 'lock-in' processes, those actors being tightly linked to the regime. This lock-in is entirely technical (actors have invested in costly specialist equipment which cannot readily be converted to other uses) and economic (actors depend on economic chains involving up-stream and downstream partners from the regime, who guarantee a significant economic return, and, therefore, allow these intermediaries to afford the high costs of their equipment). Food processing industries are tightly coupled to supermarkets and streamlined to provide large amounts of standardized products. For firms processing and providing produce to restaurants, the development of local food chains would imply an increase in the number of partnerships with local farmers, which would generate a logistical problem that they are not prepared to solve. Some local initiatives have been experimenting (such as the project 'Prefer Local', designed to organize a local chain providing meat for local butchers) but they have stopped short as soon as the subsidies supporting their development were removed.

In Santorini, the farmers initially resisted because the technical changes expected of them had a strong impact on their working conditions and their individual and collective identity. Harvesting 1 month earlier, in August rather than in September, requires the dedication of family labour to farming whilst the tourism season is at its peak. Most farmers rely on multiple small income streams based on wine and tourism activities. Farmers also objected to changing vine cultivation practices, as the new practices have changed the unique vineyard landscape of Santorini. Furthermore, most farmers were older, often cultivating vines for traditional and/or sentimental reasons, rather than purely commercial rationales.

Overcoming resistance

In Rennes, the resistance of intermediary actors to become involved in local short supply chains was partially overcome thanks to various, sometimes contradictory, dynamics. During the first stages of niche existence, farmers managed to do without the intermediaries (the latter representing 'the missing actor') and even developed various strategies to avoid them. The idea was mainly to integrate processing activities on-farm, for two reasons: first, from an economic perspective, it allowed farmers to retain a larger amount of the added value on the farm; second, from an institutional perspective (closely connected to the first point) farmers understood the role of intermediaries as capturing as much of the value added as possible in the food chain, cementing regime dynamics in their favour. Partnerships were neither needed nor desired, mainly because few intermediaries shared the niche's perspective of quality and autonomy of production.

More recently, this dynamic has evolved with a new generation of local food chain projects which have integrated intermediaries. This dynamic was driven by larger volumes of produce and higher levels of consumer demand, illustrating the strength and importance of the 'niche tandem' which reinforces transition dynamics. A new type of intermediary actor was created and integrated into the niche dynamic: butchers, logistics, restaurants and cooks either changed their practices or appeared as newcomers with innovative practices. All now understand their activity as a necessary and innovative vector of connection between producers and consumers. Different drivers explain their new attitudes. Some actors are 'generated' by the niche, such as small collective abattoirs managed either by farmers acting alone, or in partnership with a butcher. Other actors had previously existed but sought to escape from the regime, not least due to economic difficulties. This was the case of an abattoir in Rostrenen which was so pressured by down-stream partners of the regime that it went through bankruptcy before radically re-orienting its activities towards regional food chains.

In Rennes the overall result has been a change in the perspective of the niche in the last 4 years: the idea is slowly switching from a perspective based on short supply chains (which are widespread in France, Kneafsey *et al.*, 2013 and based on either no intermediaries, or only one intermediary, between producer and consumer) to a more regional vision of food chains. Produce, now expected to transit from producer to consumer at the local-regional level, has to be identified as regionally produced and as meeting the holistic quality standards which are the foundation of the niche. However, the role of, and need for, intermediaries to optimize such local food chains is slowly being recognized. These intermediaries are expected to ease logistical issues, to reduce the food miles of local food chains and help cope with the larger amounts of produce needed given increasing consumer demand. One of the indicators of this evolution is a new research programme funded by the French Ministry of Agriculture, called 'Interval', studying the characteristics and specific benefits provided by these new niche actors.

In Santorini, the resistance of farmers has been partially overcome thanks to two simultaneous processes. The first process initiated by niche actors was based on a 'semi-forced' enrolment of existing grape producers by increasing the prices paid for grapes and offering technical assistance. Initially, the prices paid to farmers were suddenly doubled by winemakers in the niche who had individually settled as new entrants. Later, Santo Wines also followed this example by increasing (though not doubling) the price paid to farmers.

The second process consisted of enrolment of new actors (in this case new wine-makers) representing the 'missing' part of the niche, and meeting increasing demand. Some niche actors, pioneers and more innovative, well-established winemakers, brought changes in production practices such as linear systems of vine cultivation despite resistance from the regime. Older farmers were more reluctant to adopt such changes because of deeply held visions of farming and farmer identities. By incorporating the niche's existing technology and marketing practices, SantoWines has become bigger and more modern, and gained more power and influence over farmers and local society. The fact that SantoWines is the compulsory last resort buyer of local grapes gives it the power to formally represent local farmers. Senior executives of SantoWines have also gained power in local politics. For example, the Chairman of SantoWines is an elected local politician. Lastly, SantoWines now also handles all subsidies for farmers, another example of its power base.

Conclusion

In Rennes and Santorini (and possibly in Pilsen, in the future trajectory of the niche), the enrolment of resistant actors into the niche is based on their progressive interest in the activity of the niche. In Rennes, intermediary food chain actors joining the niche benefit strongly from their new economic and institutional independence towards the regime. Price negotiations with up- and down-stream actors, social recognition of their role, and social networking are strongly reinforced. The increasing presence of these actors will probably contribute in a decisive way to reinforce and stabilize the autonomy of the niche, thanks to the involvement of the full range of food chain actors. In Santorini, farmers involved in the niche earned twice the price for their grapes, were involved in a more reliable chain, as well as involved in a renewed image of the quality of their product (even though the landscape and quality issues remain factors of disagreement).

These trajectories of niches based on AAFNs fits with what Smith and Raven (2012) defined as 'stretch and transform empowerment' (processes that restructure mainstream selection environments in ways favourable to the niche). According to the authors, this progressive empowerment process, which occurs in multi-dimensional 'protected spaces', allows for the next steps of the transition, characterized by the emergence of an 'innovation-specific proto-regime' operating on two levels; local and global with local experience of socio-technical networks being converted by global networks into more generic, mobile processes and norms (Smith and Raven, 2012).

At that stage, AAFN-based innovation processes seem to have stopped following this general scheme of transition processes. The first lesson learned from our three case studies is that if empowerment in protected spaces can be observed at the local level, niche anchoring at the local level of the regime allows AAFNs to better dissociate from the anonymous macro-level sub-regime. The niche aims to achieve its autonomy from the regime, not align with it. In other words, the niche does not aim to change the rules of the regime, but rather not to depend on them anymore. The solutions used to reach this aim rely on the constitution of independent and self-sufficient AAFNs, meeting local demand whilst ensuring reasonable income to the actors (production, processing and marketing) involved in the niche.

The examples of Rennes and Santorini inspire a broader observation: when directing energy towards its development and when coping with its success, the niche needs to obtain the commitment of all actors involved in the food chain, in other words, actors representing all of the sub-regimes in the food regime. If one of the categories of actors (one of the sub-regimes) is missing, the linking process of the niche appears blocked. The transition in Pilsen is too recent to allow for such an observation, and its future trajectory will be interesting to follow for this reason alone. The commitment of all sub-regimes seems to be a necessary condition for successful anchoring of the niche to the local-level of the regime.

The second lesson learned from our three case studies is that the growing size of the niche provides the means to overcome resistance: the more developed the niche becomes, the more it is able to generate and organize solutions. We have shown that these solutions can be based on the semi-forced participation of resisting actors, on a spontaneous movement of those actors who initially resisted (for example when actors become more conscious of the benefits that involvement in the niche can provide), or on the creation of new actors who fill the gap left by the absence of resistant actors.

The third lesson learned is that actors who join the niche from resisting sub-regimes appear satisfied because they benefit from the niche's autonomy from the regime. In return, thanks to the involvement of representatives of all sub-regimes, the niche reaches a more systematic structure and dynamic and thus reinforces the means of that autonomy. Both newly enrolled actors and those enrolling them, gain from this common dynamic. Indeed, the rules within the niche are permanently re-negotiated to ensure its key values, such as negotiated prices, retention of added value within the food chain and good working conditions. The development of a strategy of autonomy and the improvement of social, technical and economic conditions in the niche are possible because everything happens at the regional level. Thus, consultations and direct negotiations are possible within the network of local actors, a process which would not be possible at the macro-level of the food regime. Autonomy and quality, both components of the framework niches based on AAFNs, are inseparable from the regional dimension of such a transition, and can be understood as its key characteristics.

References

Balasz, B. (2009) *Comparative Analysis of the Context of AAFNs at the Local, National and European Level.* FAAN Project report, European Community's Seventh Framework Programme.

Colman, T. (2008) *Wine Politics: How Governments, Environmentalists, Mobsters, and Critics Influence the Wines We Drink.* University of California Press, Berkeley, California.

Darnhofer, I. (2015) Socio-technical transitions in farming. Key concepts. In: Sutherland, L-A., Darnhofer, I., Wilson, G.A. and Zagata, L. (eds) *Transition Pathways towards Sustainability in Agriculture: Case Studies from Europe.* CABI, Wallingford, UK, pp. 17–32.

Darrot, C. and Durand, D. (2011) Référentiel central des circuits courts de proximité: Mise en évidence et statut pour l'action. In: Traversac, J.B. (ed.) *Circuits Courts, Contribution au Développement Régional.* Educagri, Dijon, France.

Elzen, B., Barbier, M., Cerf, M. and Grin, J. (2012) Stimulating transitions towards sustainable farming systems. In: Darnhofer, I., Gibbon, D. and Dedieu, B. (eds) *Farming Systems Research into the 21st century: The New Dynamic.* Springer, Dordrecht, the Netherlands, pp. 431-455.

Gale, F. (1997) Direct farm marketing as a rural development tool. *Rural Development Perspectives* 2, 19–25.

Geels, F.W. (2010) Ontologies, socio-technical transitions (to sustainability), and the multi-level perspective. *Research Policy* 39, 495–510.

Kneafsey, M., Venn, L., Schmutz, U., Balázs, B., Trenchard, L., Eyden-Wood, T., Bos, E., Sutton, G. and Blackett, M. (2013) *Short Food Supply Chains and Local Food Systems in the EU. A State of Play of Their Socio-economic Characteristics.* European Commission Joint Research Centre, Institute for Prospective Technological Studies, Seville, Spain.

Maréchal, G. (ed.) (2008) *Les Circuits Courts Alimentaires: Bien Manger dans les Territoires.* Educagri, Dijon, France.

Marsden, T.K. (2006) The road towards sustainable rural development: Issues of theory, policy and practice in a European context. In: Cloke, P.J., Marsden, T. and Mooney, P.H. (eds) *Handbook of Rural Studies.* Sage, London, UK, pp. 201–213.

Marsden, T., Banks, J. and Bristow, G. (2000) Food supply chain approaches: Exploring their role in rural development. *Sociologia Ruralis* 40, 424–438.

Murdoch, J., Marsden, T. and Banks, J. (2000) Quality, nature, and embeddedness: Some theoretical considerations in the context of the food sector. *Economic Geography* 76, 107–125.

Renting, H., Marsden, T. and Banks, J. (2003) Understanding alternative food networks: Exploring the role of short food supply chains in rural development. *Environment and Planning A* 35, 393–411.

Smith, A. and Raven, B. (2012) What is protective space? Reconsidering niches transitions to sustainability. *Research Policy* 4, 1025–1036.

Spaargaren, G., Oosterveer, P.J.M. and Loeber, A.M.C. (2012) Sustainability transitions in food consumption, retail and production. In: Spaargaren, G., Oosterveer, P.J.M. and Loeber, A.M.C. (eds) *Food Practices in Transition – Changing Food Consumption, Retail and Production in the Age of Reflexive Modernity.* Routledge Studies in Sustainability Transitions. Routledge, New York, pp. 1–31.

Verhaegen, E. (2012) Les réseaux agro-alimentaires alternatifs: Transformations globales ou nouvelle segmentation du marché ? In: van Dam, D., Streith, M., Nizet, J. and Stassart, P.M. (eds) *Agroécologie, Entre Pratiques et Sciences Sociales.* Educagri, Dijon, France.

Watts, D., Ilbery, B. and Maye, D. (2005) Making re-connections in agro-food geography: Alternative systems of food provision. *Progress in Human Geography* 29, 22–40.

Watts, D., Ilbery, B. and Maye, D. (2008) Making re-connections in agro-food geography: Alternative systems of food provision. In: Munton, R. (ed.) *The Rural: Critical Essays in Human Geography*. Ashgate, Farnham, UK, pp. 165–183.

Chapter 11

Local quality and certification schemes as new forms of governance in sustainability transitions

M. Lošťák[1], P. Karanikolas[2], M. Draganova[3], L. Zagata[1]

[1]*Czech University of Life Sciences, Prague (lostak@pef.czu.cz);* [2]*Agricultural University of Athens;* [3] *Institute for the Study of Societies and Knowledge, Sofia*

Introduction

Each of the 21 case studies discussed in this book highlights, to a greater or lesser extent, the issue of governance, referring primarily to the practices of a 'governing body' or 'governing principles'. The interactions between emerging niche innovations and governance arrangements are demonstrated within the case studies. These include the impacts of niche anchoring on governance (for example the establishment of new regulatory mechanisms and subsidies for renewable energy production, Sutherland *et al.*, this volume) and the influence of regime/landscape governance arrangements on the niche (such as in the case of lifestyle land management, Pinto-Correia *et al.*a, this volume). In this chapter, the changes to governance practices are the primary focus of the analysis. We demonstrate how governance arrangements are constructed by niche actors, in order to achieve goals such as the sustainability of agriculture, or rural development, utilizing local quality and certification schemes.

In this chapter, governance structures are defined as rules for organizing activities in a certain context (Swedberg, 2003). Changes to governance structures result from the reciprocal interactions of actors, agencies and structures. This is not a new principle in understanding transition processes (Kabele, 2005) or indeed social processes more generally, but the emphasis here is on how these changes occur on multiple levels. The multi-level perspective (MLP) underpinning the research in this book, offers the opportunity to assess agent-structure relationships at micro-, meso- and macro-scales, from local to global social processes. We focus, particularly, on the interactions between new forms of horizontal governance emerging in local quality and certification 'niches', and the vertical forms of governance characterizing the agriculture and tourism or environmental protection regimes in which these niches develop.

Conceptualising governance within the multi-level perspective

The need for new forms of governance in agriculture is evident in commentaries on globalization. Globalization calls for "a new, de-territorialized, cosmopolitan perspective on the (state) regulation of the global food-scape" (Spaargaren *et al.*, 2012:23). Evidence for emerging forms of market-based governance structures can be also found in Spaargaren *et al.* (2012), for example in political consumerism, when consumers vote with their wallets by buying sustainable food. Beyond these issues, this chapter will also examine other types of governance, including territorialized perspectives on regulating local activities: issues which are not necessarily demonstrated through economic behaviour (including market activities) but rather through cultural or environmental values and other non-economic activities.

The chapter will contribute to the discourse on governance in transitions in the agri-food regime. From the multi-level perspective, the regime orients and coordinates the activities of actors and, as such, accounts for the stability of the relevant socio-technical system (Geels, 2004). If the dominant regime displays significant problems, a 'window of opportunity' opens in which existing novelties that address those processes have the opportunity to influence the regime and, if successful, constitute a transition (Geels, 2005). These novelties are generated in niches, whilst incremental innovations are developed in regimes (Geels, 2004). A further important level is the socio-technical landscape consisting of a set of deep structural trends which impact both regime and niches (Spaargaren *et al.*, 2012).

The ideas presented in this chapter reflect global–local dimensions of agricultural activities and the dilemma concerning the factors of social change: is the dominant influence of modern social development shaped by economic (market) or non-economic (cultural) factors (Giddens, 1989)? In our approach we draw on the work of Marsden (2013), who suggests that when dealing with transitions in agri-food systems, two distinctive features need to be taken into account: the spatial embeddedness of the system, and the styles of governance and regulation (due to the high level of government intervention in agri-food systems) conditioning the transition. Marsden (2013) demonstrates the power of reflexive governance induced by landscape pressures which coincides with epistemic and scientific articulation, reinforcing potentials of alternative niches. We also draw on Brunori *et al.* (2012:27) in our arguments that entrainment of innovations into the system depends not only on the reflexivity of governance structures "but also on the capacity of alternative networks to adopt the evolutionary approach," by looking at the consolidation of new patterns and expanding innovative activities into other fields. We show that food oriented consumer networks "can easily move to other consumption goods" (Brunori *et al.* 2012:27). This can be understood as the role of networks in bridging various regimes, another important aspect of the analysis of our case studies, as it suggests that regimes influence each other through 'regime graffiting' (described below).

Developing Kuhn's (1970) thoughts about scientific revolutions in the MLP discourse, we started by considering situations where actors faced problems, anomalies or crises because the regime could not cope with novelty. Such situations call for radical innovations. As a result "the proliferation of competing articulations, the willingness to try anything, the expression of explicit discontent, the recourse to philosophy and to debate over fundamentals, all these are symptoms of a transition from normal to extraordinary research" (Kuhn 1970:91). However, competition from constructed novelties raises the question about their success in terms of their sustainability. Drawing on examples from science, if scientists are to challenge the dominant paradigm, they must reflect on existing values and norms and find new ways to see the world. Sometimes, it is easier to draw on the paradigms of other sciences through paradigmatic graffiting (Stepin, 2005). We will use these thoughts to document a type of 'regime graffiting' that we found in our case studies. This term means that some of the principles which construct governance arrangements existing in one regime can 'overwrite', through social networks and niches, the governance arrangements in another regime, in the sense of making a new form. The methods of such 'overwriting' depend on how actors reflect regime governance structures in constructing their niches, and how well supported they are by wider societal trends.

The overarching question which frames this chapter is: how is the development of each of the three case studies related to the development of new governance structures? Because governance structures emerge as a result of the mobilization of resources through social networks, often against the constraints of previous development (Swedberg and Granovetter, 1992), we will also look at how they are constructed in the agriculture and rural development sectors through networking, in spite of locked-in practices implemented in the dominant regime. To achieve this goal, we will assess why some niches are successful and others not, demonstrating that appropriate combinations of, and links between, social and cultural elements in niches, regimes and landscapes; networks of actors (including their cultural and social backgrounds); and governance structures are highly important for transitions to emerge and succeed. We also assess the role of EU policy, both in enabling and in constraining niche development in different settings.

Description of the cases

The case studies in this chapter are situated in Bulgaria, the Czech Republic and Greece (Fig. 11.1). All three cases are located in mountainous areas and operate under environmental regulations. The economies of the study areas are dominated by small and medium-sized businesses which are active in food-processing, processing of non-food resources typical for the region (such as wood and textiles) and in the tourism and hospitality sectors. Agriculture still plays an important role (although it is not dominant) in all three regional economies. The different temporal starting points of the niches do not allow for the establishment of a coherent time line bringing all three cases together, but the comparison enables the assessment of similar processes at different stages of development.

The Bulgarian niche (initiated in the early 2000s) is located in the municipality of Elena (central northern part of Bulgaria). The initiative was established by an NGO (Local Tourism Council) in cooperation with local municipal authorities and other local tourism or food processing, as well as other local NGOs and actors. NGOs are the main organizations through which local people are involved. The goal of the initiative is to achieve sustainable development at municipal level, through creating tourist products which will integrate rural tourism, traditional agriculture and food production into a coherent set of offers for tourists. Progress is evident in the growing number of tourists, the establishment of organic bee-keeping in the area and an increase in the amount of land planted with fruit trees. Joining rural tourism with agriculture and food processing in 2005, local actors started working on the establishment of a certification scheme for local products (e.g. organic honey, Elenski ham, perennial plants). This in turn resulted in the establishment of a public-private structure (Municipal Expert Commission) for the categorization of accommodation facilities, communal labelling and direct marketing of organic honey, and support for the process to protect and brand local ham. This scheme is embedded in a new governance structure; the development of a close partnership between the actors involved, and a public forum to develop the strategies for local tourism and municipal development through the Local Action Group in a LEADER-style approach (LEADER is a bottom-up approach to rural development in Europe, see European Commission, 2000). The scheme is expanding the involvement of consumers in the validation of new products and services. The actors anticipate that it will help to popularize local amenities and promote the municipality as a rural tourism location. The initiative cannot be said to have already completed its goal because the tangible evidence of success is still fairly limited: the long term commitment of all actors is still unclear; communication between them is sometimes insufficient and unprofessional; and the initiative is not yet attracting the desired numbers of actors and tourists. National regulations related to direct sales from farms also block some of the strategies. Therefore, participants have expressed hope that the local certification scheme, if well developed and supported will be more successful in helping them to achieve their goals.

The Czech niche (established in the early 1990s), situated in the region of the White Carpathians Mountains (south-east border of the Czech Republic), also aims to capitalize on the region's specific natural and cultural heritage to support local sustainable development. The initiative introduced a regional label in late 1990s which certifies high quality products that uniquely represent local traditions. The label was introduced about 10 years before similar activities were developed at the national level. The first steps of the initiative can be traced back to the early 1990s when older owners of fruit orchards met younger members of a local environmental protection organization. Together, they recognized that existing farming and forestry practices, supported by centralized and productivist governance structures, were leading to the ongoing loss of old fruit trees, which had traditionally characterized the local landscape. They decided to map out those 'heritage' varieties and protect them for the future. In the mid-1990s, organic farmers joined the initiative. This cooperation worked on an informal basis until the late 1990s when the participants established a formal NGO – the civic association 'Tradice Bílých Karpat' (Tradition of White Carpathians) – which enabled more intensive participation by activists from the city of Brno, whose focus was the theory and practice of sustainability.

There were two main reasons for the shift into formal cooperation. One was linked to strategic vision: to go beyond mapping the old trees and to use the heritage of the region for tangible evidence of development (such as through processing apples into cider, or drying fruits using heritage technologies). The second reason was pragmatic: to achieve the first goal they needed to apply for a grant from Luxembourg which was only accessible to legal entities. The change into a legal entity also resulted in the formation of a local certification scheme and regional label, which is awarded to selected local products from the region. The money generated through activities and the certification scheme (for example through the sale of organic apple cider) is intended to support new projects (such as micro-financing) which will increase the sustainability of the region and will also generate funds for similar activities. At the time of the study in 2012, the label listed about 20 such producers and their products ranged from apple cider to honey, decorated gingerbread, lamb, liquors, ceramics and lace. However, although the work of the initiative has gained a considerable reputation (it was acknowledged by a visit from Prince Charles in 2010, who met with certified producers), issues continue to emerge. There is currently some conflict between organic farmers who pursue economic profit, and environmentalists who are often newcomers to the region, who emphasize nature conservation values. Linking local certification with new national schemes, which are now heavily supported by the state, is also a challenge; as is how to sell the products, as retail chains are unsupportive. The initiative also faces problems accessing the glass bottle deposit scheme controlled by large beverage producers. At present, bottles are not returned to cider producers, making it less competitive than more commercial operations. Local people have also become less and less active in activities undertaken in the region and those who are not involved have developed a rhetoric that the initiative is 'working only for itself' in the sense of operating solely to benefit its members.

The Greek niche (initiated in the late 1980s) operating in the Plastiras Lake area of the Karditsa region (a mountainous area in central Greece) introduced an organizational innovation into the local economy through a new form of governance related to the 'Local Quality Convention'. The formal goal of the Convention included upgrading income and improving living standards through the protection of the environment and strengthening of local tourism services. The Convention came about as a response to challenges faced in the area (few employment opportunities for local people) through the active involvement of a range of actors (often the descendants of the original inhabitants who had recently moved back to the locality, bringing with them their experiences from other businesses) and a set of obligations and rules for participating enterprises. The scheme concerned aesthetic standards for various facilities related to rural tourism (information signs, kiosks for selling local products and also conservation and promotion of culinary heritage using local products). The key actor in the initiative was the ANKA development agency, established in 1989, which brought together various actors and organizations (e.g. the Union of Agricultural Cooperatives or Local Union of Municipalities) operating in the region. The agency is a member of multiple networks (with universities, research centres and other development agencies) and is, thus, important for networking and providing information to those 'in need' in the locality. Other important networks are those of new entrepreneurs (non-rural individuals who come back to the area of their origin) which span extended extra-local networks.

From 1999, the initiative aimed to include quality in all aspects of the local economy through a process assured by a certification scheme. Koutsouris (2009) has outlined the method of social and spatial deliberation developed by ANKA for the formation of the Local Quality Convention. The phases of development reflect the problems experienced by the Convention. Although it had previously been considered an example of success, because its ideas were adopted by 11 regional development agencies in the Pindos Mountains at the time that LEADER 1 ended, by the mid-2000s, the initiative had lost internal coherence and strength. This loss of coherence became obvious when opportunistic behaviour emerged (in the form of personal and parochial political attacks on the first president of the Convention). These problems continued to derail the purpose of the initiative and it became vulnerable. It was not able to foster sufficient collective action in the area, as it became an instrumental arrangement preoccupied with its members' interests and dominated by new entrepreneurs. The three cases are discussed further below.

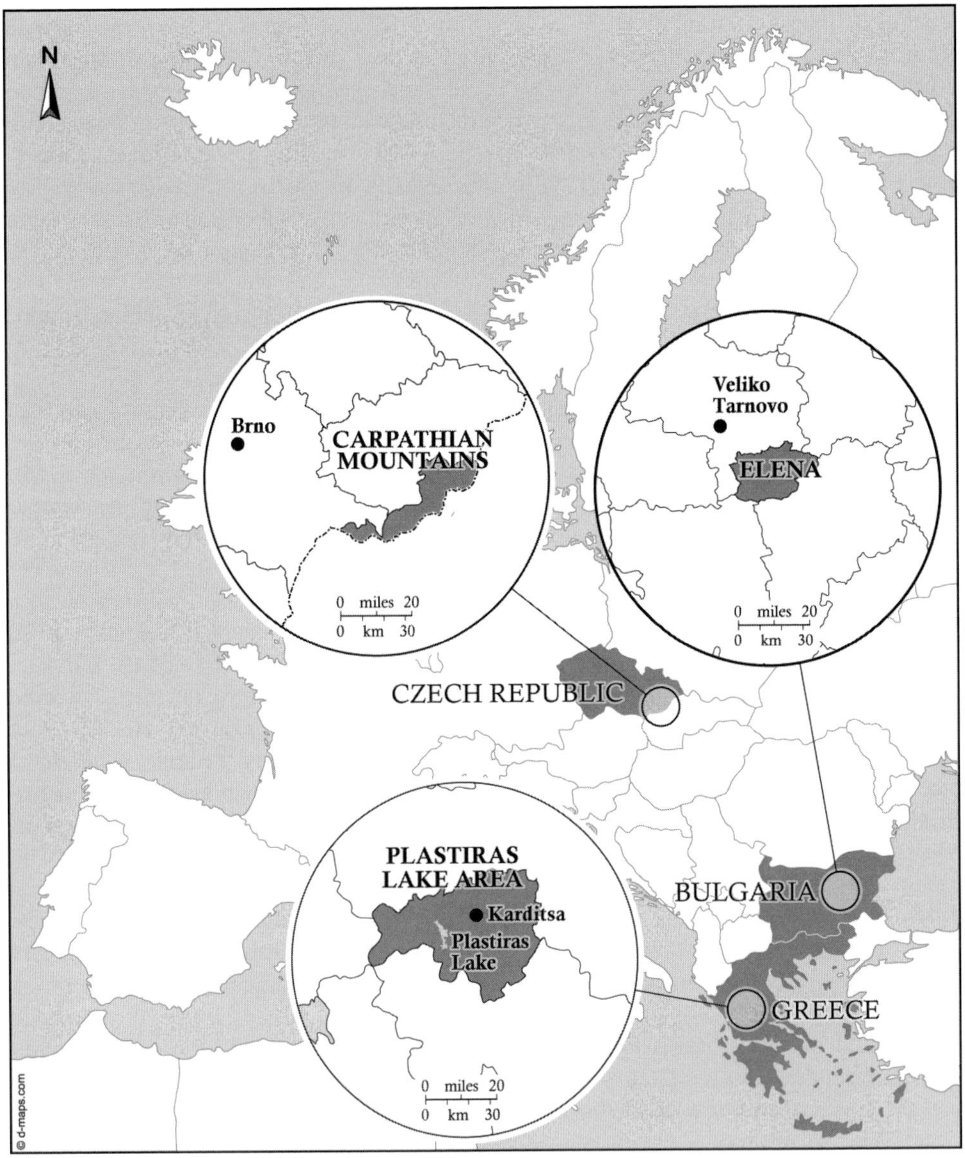

Fig. 11.1. Location of the three case studies: Elena, Carpathian Mountains and Plastiras Lake (Source: authors).

Lessons learned from the case studies

The initiatives in all three case studies address the sustainability, flexibility and efficiency of forms of governance in the regime, although they each started at different times and under different landscape pressures (Common Agricultural Policy (CAP) reforms in Greece; the collapse of neo-Stalinist collectivization, Swain, 1998 in the Czech Republic and Bulgaria; together with the EU accession of both countries but at different times). From the perspective of niche actors, the regime's governance structures face the challenge of dealing with specific problems in the localities where they operate. These problems include depopulation, growing unemployment, decreasing numbers and variety of species (e.g. fruit trees), lack of authenticity in food products, and the de-coupling of traditional human–nature relations. Niche actors consider regime governance structures as responsible for the apathy and inactivity of the target population (mostly in Greece and Bulgaria) and also for damaging human–nature relations (in the Czech Republic). The new forms of governance emerging through niches therefore represent direct or indirect responses to regime dysfunction. The new structures represent horizontal networks of actors, including farmers, food producers and processors, environmentalists, consumers and tourists, in order to engage the local population or improve human–nature relations to address the problems they are facing in a more efficient and sustainable manner. Locality, quality, networks (cooperation) and traditions were the words most used in discussions about local certification schemes and organizational innovations, which were developed to counter regime tendencies towards globalization, quantity and material growth. The schemes were not developed as new forms of governance quickly but took 5 to 10 years after the establishment of the initiatives to develop.

The landscape-regime-niche interactions indicate an amalgamation of vertical (typical for modern organizations) and horizontal (late modern fluid networks) forms of governance. For instance, the Czech case was significantly developed through the use of an international grant (vertically channelled funding) which was used to reconstruct an old barn for the production of local organic apple cider. The grant was the result of international networking by environmentalists participating in the initiative. Later, the actors in the network learned that to market the cider it had to differ from other beverages; they therefore developed the first local label and extended the local certification scheme to other quality small-scale products in the area. The network now utilizes the money generated through the scheme to support other projects with similar aims.

The emergence of the forms of governance described in this chapter was influenced and supported by the socio-technical landscape through its values and norms, particularly CAP reforms, including LEADER approaches and the decentralization discourse. In the Czech Republic and Bulgaria, this process was also related to the collapse of centralized forms of national and local governance of the economy and society. Contemporary landscape pressures were also related to social trends such as consumer preferences towards food quality, and tracing authenticity of products or activities. Environmental concerns vis-à-vis economic and market dominant discourses also played an important role. These niches focus on decentralization, including endogenous approaches to rural development, local food and services quality and high-level environmental protection, as well as the maintenance of traditional knowledge and products.

These niches challenge the technological aspects of the regime through traditional and local production techniques in food processing (also environmental protection and traditional farming methods in the Czech case), off-farm and non-farm activities (tourism and services), in contrast to the sectoral focus of farming. In the Czech case, there is also significant orientation towards the use of green technologies, such as an efficient community heating system using renewable resources (wood from community forests) or fuelled by small community-owned solar energy stations, and the use of wetlands to treat waste water. From this perspective, the niches also highlight the use of 'retro-innovative' technologies: traditional practices which are ‘reinvented’ as novelties compared with practices or products generated by the regime (e.g. using a 250 year old kiln to dry fruits in the Czech Republic or developing traditional crafts and foods in Greece and Bulgaria). These initiatives demonstrate a willingness to experiment with traditional practices which are often, but not always, fashionable in modern society. The niches protect these retro-innovations through locally designed certification and quality assurance schemes.

All three initiatives have fostered cooperation amongst local actors. This is an important change compared to the limited amount of collective action, and the apathy and individualism of actors prior to the start of the initiatives. Collective action necessitates networking, which is not primarily oriented to agriculture alone but is more multifunctional, addressing social, environmental and economic goals. The bridges between networks are important for the initiatives to work. They enabled, at least at the beginning, the transfer of new ideas from outside to influence and help the initiatives to develop.

Key to these new networks were newcomers who came to the area (and to farming) with new ideas. As such, they represent reflexive actors; as the initiators of their respective initiatives, they were the ‘engines of change’. They tended to have different backgrounds to the majority of the local population, in that although they were, in many cases, the descendants of local people, they had lived in urban areas. They recognized the potential of the lake in the Greek case for irrigation and also for tourism. In the Czech case they saw sustainability in its three dimensions: processing organic apples to produce quality cider which provided local jobs, was profitable and beneficial to the local environment. In the Bulgarian case, they developed ideas about integrating rural tourism and the value of nature protection.

Although the newcomers were important in the early part of the process, once the initiatives had established, the new forms of governance, surprisingly, hindered the transition process. In Greece and the Czech Republic (Bulgaria is at an earlier stage of transition) the niches and their networks, although successful, are seen as operating primarily to serve their members. As such, they seem closed and are not generating the expected benefit for the whole region. The Greek case is both the oldest and the one where this problem is the most critical for the future development of the niche. In the Czech Republic, local inhabitants take the initiative for granted and are much less active than they were in the past. Similar to the Greek case, the activities centre on the core members of the network. Niche networks therefore may face resistance when anchoring into regime networks (actors who are not members of the ‘core niche network’), impacting negatively on the niche.

The outcomes of the reflexive governance structures are also supported by the existence of similar types of initiatives elsewhere. The Greek initiative was impacted most by the existence of a local LEADER project, albeit indirectly: locals considered the key actor of the initiative (the development agency ANKA) and the LEADER project to be synonymous, with new entrepreneurs coming to the area and therefore believing that LEADER was not for them. On the other hand, the Bulgarian case was strongly influenced by contemporary EU policy and its measures and less by LEADER (the Local Action Group did not submit a local development strategy for funding). The Czech case was less influenced by EU rural development policy measures, although the actors in this case learned from examples outside the Czech Republic through the international networks of environmentalists. This case implements the principles of LEADER without being strongly committed to the LEADER Local Action Group. Some of LEADER principles were used even before this approach was launched in the Czech Republic due to international networking and reflexive learning.

Rethinking transition in forms of governance

Although the actors in the case studies challenged the governance structures of the dominant regime, they only partly succeeded in counterbalancing the emerging problems. The longer term, originally successful, Greek and Czech cases (compared to the Bulgarian case) now evidence problems linked with a change from strong-weak networks into weak-strong networks (Granovetter, 1973). That is, ties outside the niche that were important for anchoring (weak in terms of familiarity among members of various networks but strong in terms of access to information) have become inwardly focused and limiting (closure of networks resulting from close familiarity of network members results in weak access to new information). Local certification is embedded in locality; the initiatives studied appear to be closed and not expanding. They face difficulties when anchoring local niche networks into the national regime. On the other hand, under appropriate landscape pressures (for example consumers' calls for food authenticity in the Czech case or increased demands for alternative tourism in the Bulgarian and Greek cases) the initiative might influence the expansion of such local labels at the national scale.

Similar to many other forms of local quality initiatives across Europe (Eden *et al.*, 2008) and indeed across the world (González and Nigh, 2005), the three cases offer guaranteed local quality and information for consumers instead of focusing on quantity from depersonalized 'nowhere' locations of products from the agri-food regime. The agri-food regime referred to in the case studies was constructed in accordance with the values and norms of modernity. These were based on the belief that positive possibilities discovered in the modern era will prevail over negative features of the modern period (Giddens, 1990). The emerging problems faced by the regime led participants to question its sustainability – sometimes tacitly (Greek and Bulgarian cases), sometime intentionally (Czech case). The key questions raised were: how is the regime to be sustainable when it contributes to inaction, creates environmental problems, eliminates variety, brings unemployment and results in depopulation? As a reaction to these problems, a proliferation of actions was evident. They concerned amenity provision, environmental protection, local food, food quality, traditions and tourism.

The cases studied did not change agri-food regime governance structures at national level. Instead, the niches form 'islands of positive deviations' at regional level – a space for independent thinking and acting (Bútora, 2006). In this way, they significantly differ from regime governance structures, although the actors in the niche must also operate within some aspects of the regime (e.g. collection and re-use of cider bottles in the Czech case or regulations on food processing in the Bulgarian case).

The analysis of the niches shows the transition might be more likely to occur when two or more sub-regimes are closer in terms of the influence they exert on each other. This was the case for the agri-food regime and tourist regime in Bulgaria; the agri-food, tourist and also energy regimes in Greece, where a large lake was being used for energy production; and the environmental protection regime and agri-food regime in the Czech Republic. This overlapping of regimes creates the opportunity to bring various views together into a new niche, with associated novel governance structures. Using the concept of 'paradigmatic graffiting' (Stepin, 2005) we will consider relations between regimes and niches: when one paradigm can 'overwrite' the other and shift the paradigm if this is socially and culturally backed and reflected through philosophy. In order to construct new governance structures in niches in the agri-food regime, either reflection of existing governance structures in a tourist regime, or environmental protection regimes were used in order to help with 'regime graffiting'. Some elements of the other governance structures were taken into the niche developed in the agri-food regime and later 're-wrote' the original governance structures, orienting them towards the importance of pioneering organizational innovation. This process was particularly evident in Greece, where the case influenced regional governance structures and in the Czech Republic, where there is now a boom in regional labelling, and national bodies are working to coordinate the initiatives under one umbrella.

Networking, in all three cases, reflects late modern discourses about the fluidity of contemporary society (Bauman, 2000). New ideas, brought by those who are often not embedded in the agri-food regime, are backed by broader landscape changes. The agri-food regime is typified by hierarchical forms of governance, whereas the niches studied have developed horizontal networks to establish their own governance structures. At the beginning of the initiatives these networks were fluid, later (5 to 10 years) they were protected by local schemes. Networks and forms of governance were seen as the tool to activate people, to make the agri-food regime more efficient, more quality oriented, or to raise public awareness towards the environment. 'Regime graffiting' enables opening the regime to actors with alternative viewpoints, due to the capacity of the networks; for example, consumers (in all three cases) and environmentalists (in the Czech case) became important actors in the agri-food regime when governance structures in the niches were changed to provide flexible networking and were protected by certification schemes.

However, such developments are not always beneficial for the initiative in its later stages, where they may slow down its drive. In the Czech case, there were disagreements between farmers concerned with profits, and environmentalists for whom the value of environmental protection and the traditional cultural landscape was more important. Such contradictory objectives in the network had negative impacts on the perception of the niche by both members and non-members. A similar situation was found in the Bulgarian case where the profit-seeking of hotel owners weakened the links in the network which were needed to achieve integrated tourism products to benefit the whole community. In Greece, the niche restricted its activities primarily to the members of the core network and, as a result, the initiative did not facilitate a reflexive process or the engagement of new stakeholders in joint learning and action. Despite these issues, network members indicated that incorporating more actors into niches through networks did give the niches increased backing in society; the niche was not seen as the arena inhabited solely by farmers. The inclusion of non-farming actors results in broader impacts on society and appears to be more sustainable because diverse actors indicate a more socially accepted initiative.

An important element in the new governance structures was the use of some sort of local certification and quality scheme, which countered the governance structures of the regime. These schemes are inherently local and, therefore, spatially specific. These findings are consistent with Marsden's (2013) ideas about the importance of two distinctive features of transition in agri-food regime: spatial embeddedness; and high levels of state intervention. These schemes, as well as the governance structures, were linked with networks. These networks and their local quality and certification schemes were seen by niche actors to be the tool to activate local engagement and to deal with the problems faced by the regime. However, the links between actors and the governance structures constructed by them (represented by local quality or certification schemes) are highly important for the success of the initiative. If these schemes are merely an instrument without engendering collective action, then initiatives tend towards crisis as evidenced in the Greek case.

The forms of governance in our case studies are also linked with certain retro-elements in terms of retro-technologies, retro-practices, and in general retro-innovations, as well as greater emphasis on green technologies or green practices. These elements were most obvious in the Czech case, in the orientation towards renewable sources of energy. As for retro-practices, this initiative has the word 'traditions' in its name, which it demonstrates through activities such as traditional methods of drying fruit. The importance of retro-practices can also be identified in the Bulgarian case where traditional crafts, food (e.g. dried cured ham, local plum brandy, local honey or aronia wine) and traditional agriculture are the most important elements of the niche; and in Greece, with newcomers appreciating the value of tradition combined with their outsiders' expertise. Such retro-elements do not represent retrograde transitions (Wilson, 2007) because they influence the transition to sustainable use of local resources and promote new values, patterns and quality of consumption. They are linked to new forms of governance challenging regime governance structures. Because the new forms of governance are constructed by modern actors, with their fluidity and flexibility, the technologies, in their retro-fashion, reflect Baudrillard's (1994) understanding of retro-practices as an attempt to resurrect the period when at least there were some strong and unchangeable guiding principles.

Reflecting existing governance structures, newcomers can incorporate their views and related practices into niches because the views are supported by the wider context. It is the window of opportunity which provides the values and norms needed for the new governance structures to be protected in the niche. The window of opportunity makes emerging governance structures more attractive for new entrants, such as tourism entrepreneurs or environmentalists. They are not burdened with the entrenched views which dominate the regime. The structures are created by the newcomers and attract other new entrants to farming in the locality. The newcomers emulate Weber's young men with new values (2001), going out into the country and destroying 'business as usual' by setting up new governance structures.

On the other hand, newcomers only succeed locally through their networks. The more the initiative matures, the more problems they face expanding the niche into the regime (e.g. the Greek and Czech cases, which appear to have networks that are closing down and the problems recycling bottles in the Czech case). The new values and views are questioned by those who are not part of the initiative. Even in the niches we can observe conflicts between views emphasizing the role of the market (profit-oriented activities), which echoes the dominant regime discourse, and the views supporting intangible values, such as nature amenities or social cohesion.

Conclusion

Our case studies have confirmed that new governance structures are constructed by actors networking in niches. These structures are not constructed immediately after the origin of the niche; the initiative tends to experiment in searching for a way to cope with the problems generated by the governance structures. However, in order to continue developing, the niche has to establish its own governance structures which are different from those mainstreamed in the regime. In our case studies, these were local quality and certification schemes which actors perceived as the way to protect and develop the initiative towards sustainability. These schemes can further develop into local micro-financing activities, as the Czech case demonstrates, supporting the production of sustainable local products or helping to engage local people, as in the Bulgarian and Greek cases, if the links between networks and schemes are designed to support collective action and not only as an instrumental arrangement. The reason why actors create niches as 'islands of positive deviation' is because, in their view, regime governance structures are not able to deal with the problems in the regime. They construct new governance structures in parallel to the regime, without the ambition to change it, because these actors are locally embedded. Many of them are often not specifically related to the agri-food regime when they join. They are newcomers, with views and experiences from other regimes. As a result, corresponding values and norms enable transitions. If the initiative is successful, and is used by others as an example of good practice, then the governance structures are supported by regime actors, as was the case in the Czech 'Regional Food' project which was adopted by the Ministry of Agriculture.

Landscape, as a set of deep structural trends, values and norms, is a highly important element in supporting the transition processes. It is the landscape which provides actors with support for their reflexivity on the views existing in the regime, in order to create a niche governance structure. Landscape changes can open windows for 'regime graffiting' when actors such as newcomers from other regimes (tourism and environment protection in our cases), start to operate in the agri-food regime and bring with them their ideas (often contradicting existing views) from other regimes.

Although the networks seem to be of late modern, fluid and liquid design, the actors solidify them through references to traditions, or embeddedness in the locality, when developing new forms of governance. The latter echoes the nature of farming – being always somehow embedded in the locality since land cannot be easily transferred into other parts of the world. This characteristic of farming also impacts on the transitions to new forms of governance in agriculture: challenging modern governance structures from traditional perspectives and also from the late modern point of view.

References

Baudrillard, J. (1994) *Simulacra and Simulation* (translated by Sheila Faria Glaser). The University of Michigan Press, Ann Arbor, Michigan.

Bauman, Z. (2000) *Liquid Modernity*. Polity Press, Cambridge, UK.

Brunori, G., Rossi, A. and Guidi, F. (2012) On the new social relations around and beyond food. Analysing consumers' role and action in Gruppi di Acquisto Solidale (Solidarity Purchasing Group). *Sociologia Ruralis* 52, 1–30.

Bútora, M. (2006) Osobný vinš Miloslavovi Petruskovi (Personal congratulation to Miroslav Petrusek). *Sociologický časopis/Czech Sociological Review* 62, 1043–1054.

Eden, S., Bear, C. and Walker, G. (2008) Understanding and (dis)trusting food assurance schemes: Confidence and the 'knowledge fix'. *Journal of Rural Studies* 24, 1–14.

European Commission (2000) Commission notice to member states of 14 April 2000 laying down guidelines for the Community initiative for rural development (LEADER+). Available at: ec.europa.eu/agriculture/rur/leaderplus/pdf/library/methodology/

Geels, F.W. (2004) Understanding system innovations: A critical literature review and a conceptual synthesis. In: Elzen, B., Geels, F.W. and Green, K. (eds) *System Innovations and the Transition to Sustainability. Theory, Evidence and Policy*. Edward Elgar Publishing, Cheltenham, UK, pp. 19–48.

Geels, F.W. (2005) Co-evolution of technology and society: The transition in water supply and personal hygiene in the Netherlands (1850–1930) – a case study in multilevel perspective. *Technology in Society* 27, 363–397.

Giddens, A. (1989) *Sociology*, 2nd edn. Polity Press, Cambridge, UK.

Giddens, A. (1990) *The Consequences of Modernity*. Polity Press, Cambridge, UK.

González, A.A. and Nigh, R. (2005) Smallholder participation and certification of organic farm products in Mexico. *Journal of Rural Studies* 21, 449–460.

Granovetter, M. (1973) The strength of weak ties. *American Journal of Sociology* 78, 1360–1380.

Kabele, J. (2005) *Z Kapitalismu do Socialismu a Zpět. Teoretické Vyšetřování Přerodů Československa a České Republiky (From Capitalism to Socialism and Back. Theoretical Investigation of the Transitions of Czechoslovakia and the Czech Republic)*. Karolinum, Praha, Czech Republic.

Koutsouris, A. (2009) Social learning and sustainable tourism development; local quality conventions in tourism: A Greek case study. *Journal of Sustainable Tourism* 17, 567–581.

Kuhn, T.S. (1970) *The Structure of Scientific Revolutions*. University of Chicago Press, Chicago, Illinois.

Marsden, T. (2013) From post-productionism to reflexive governance: Contested transitions in securing more sustainable food futures. *Journal of Rural Studies* 29 (Special issue), 123–134.

Pinto-Correia, T., Gonzalez, C., Sutherland, L-A. and Peneva, M. (2015a) Lifestyle farming: Countryside consumption and transition towards new farming models. In: Sutherland, L-A., Darnhofer, I., Wilson, G.A. and Zagata, L. (eds) *Transition Pathways towards Sustainability in Agriculture: Case Studies from Europe*. CABI, Wallingford, UK, pp. 67–82.

Spaargaren, G., Oosterveer, P. and Loeber, A. (2012) Sustainability transitions in food consumption, retail and production. In: Spaargaren, G., Oosterveer, P. and Loeber, A. (eds) *Food Practices in Transition: Changing Food Consumption, Retail and Production in the Age of Reflexive Modernity*. Routledge, London, UK, pp. 1–31.

Stepin, V. (2005) *Theoretical Knowledge*. Synthese Library, Volume 326. Springer, Dordrecht, the Netherlands.

Sutherland, L-A., Peter, S. and Zagata, L. (2015) On-farm renewable energy: A 'classic case' of technological transition. In: Sutherland, L-A., Darnhofer, I., Wilson, G.A. and Zagata, L. (eds) *Transition Pathways towards Sustainability in Agriculture: Case Studies from Europe*. CABI, Wallingford, UK, pp. 127–142.

Swain, N. (1998) A framework for comparing social change in the post-socialist countryside. *Eastern European Countryside* 4, 5–18.

Swedberg, R. (2003) *Principles of Economic Sociology*. Princeton University Press, Princeton, New Jersey.

Swedberg, R. and Granovetter, M. (1992) Introduction. In: Granovetter, M. and Swedberg, R. (eds) *The Sociology of Economic Life*. Westview Press, Boulder, Colorado, pp. 1–36.

Weber, M. (2001) *The Protestant Ethics and the Spirit of Capitalism* (translated by Talcott Parsons). First published 1930 by Allen and Unwin. Routledge Classics, London, UK.

Wilson, G.A. (2007) *Multifunctional Agriculture. A Transition Theory Perspective*. CABI, Wallingford, UK.

Chapter 12

Transdisciplinarity in deriving sustainability pathways for agriculture

T. Pinto-Correia[1], A. McKee[2], H. Guimarães[1]

[1]*University of Évora (mptc@uevora.pt);* [2]*James Hutton Institute, Aberdeen*

Introduction

The aim of the FarmPath project has been to identify mechanisms that promote transition processes in the farming sector, and that support the adaptative and reflexive capacity of those involved at multiple scales. Transitions to sustainability refer to radical transformations towards a more sustainable society as a response to a number of persistent overarching problems (Grin *et al.*, 2010). For transition to occur, as has been shown in Schiller *et al.* and Darrot *et al.* (both this volume), different types of stakeholders at the niche level, and others at the sub-regime and regime level need to be engaged, take part in the change process and interact.

In this book we have argued that major transformational changes involve contestation of dominant paradigms and re-balancing of the main drivers affecting the agricultural sector. In order to find new pathways, Marsden (2013) argues that a more reflexive governance approach at multiple scales is required (see also Marsden *et al.*, 2010). In reflexive governance, both the adaptive and the reflexive capacity of the actors involved are strengthened by social learning, defined by Darnhofer *et al.* (2012) as the systematic learning process amongst multiple actors who, together, define a purpose related to the agreed need for concerted action at a variety of scales. Through social learning, farmers and other stakeholders become experts, instead of users or adopters of scientific recommendations (Röling and Wagemakers, 1998). Recent trends in transition management have evolved further in co-design, where knowledge is developed in a complex, interactive design process with a range of stakeholders involved through a process of *social learning* (Grin *et al.*, 2010).

Following the study of different case studies, FarmPath applied a participatory, transdisciplinary approach to identify visions for the future of agriculture and land related activities, as well as the pathways to achieve those visions. This second phase of the research was further developed in seven of the regions where the niches studied are based (with one region per study country). The aim was to involve, in each region, stakeholders and researchers in the co-construction of visions and pathways as a way to grasp the potentialities and constraints faced by agriculture regionally, and to identify mechanisms, at different levels of governance, that can support a transition towards sustainability. As

 Transition pathways towards sustainability in agriculture: Case studies from Europe (eds L-A. Sutherland *et al.*)

explained by Karanikolas *et al.* (this volume), transdisciplinarity has underpinned the project structure. The involvement of National Stakeholder Partnership Groups (NSPGs) from the beginning of the project, guiding and supporting decisions taken on project development in each region, represents a form of transdisciplinarity. Through the visioning process, stakeholder involvement was increased, moving the process further towards achieving the aim of transdisciplinary research; co-construction of research findings within a well-defined participatory approach.

Transdisciplinary research aims to transcend disciplinary limitations as well as the science-society divide, and to promote science-society dialogue (Darnhofer, this volume). It integrates potentially disparate knowledge with a view to creating useable knowledge, knowledge that can be applied in a given problem context and that has some prospect of producing change in that context (Darnhofer, this volume). In relation to transitions and their inherent transformational change, transdisciplinary approaches are advocated for use in transition management, where the multi-level perspective (MLP) is applied to action research intended to facilitate ongoing sustainability transitions. In this research process, transdisciplinary approaches were considered essential to the effective identification of sustainability visions and pathways, in order to steer change processes through the involvement of multiple actors at different levels. The aim of this chapter is, thus, to:

- Present the results of the transdisciplinary visioning process (co-constructed visions and pathways for the local study regions), highlighting the main similarities and differences between the European regions considered;
- Assess the potential outcomes and drawbacks of a transdisciplinary process in practice when dealing with transition processes in agriculture, and to reveal what appear to be the conditions for success of this type of transdisciplinary approach, in the multiple regional contexts of Europe; and
- Reflect on the complex role that the transdisciplinary dialogue had on the stakeholders involved, both the practitioners and the researchers.

The first part of this chapter reflects on how transdisciplinarity can be understood and applied in practice. The discussion then moves on to present the methodological steps taken, and provides a summary of the resulting outcomes. The chapter evaluates the transdisciplinary process, and the challenges ahead, in terms of transition processes in agriculture and rural areas.

Understanding transdisciplinarity

Transdisciplinarity is considered central to understanding and resolving sustainability challenges (Lang *et al.*, 2012). It is seen as holding much promise when tackling messy or 'wicked' social and environmental problems which are intertwined with the socio-political context and require the participation of stakeholders to generate socially acceptable outcomes (Carew and Wickson, 2010; Neef and Neubert, 2011). However, the increasing use of transdisciplinarity in political and scientific discourses, and in applied studies and research processes, has led to considerable confusion over the concept and practice (Reed, 2008). There is, therefore, an urgent need to strengthen the knowledge realm of transdisciplinary processes, and to discuss how they function and where they lead in different situations.

Transdisciplinarity is a demanding form of knowledge integration (Spangenberg, 2011), in part due to its time-consuming nature and the extent of stakeholder interaction necessary, and in part also due to its dependence on reflexivity (the process of dialogue and personal consideration that requires honesty, sincerity, and openness to diverse viewpoints and available knowledge) (Habermas, 1981; Healey, 2006; Spangenberg, 2011). Transdisciplinarity can be characterized by three elements: the integration of disciplinary paradigms; the use of participatory methods; and the application to real-life problems (Pohl and Hirsch Hadorn, 2007). It is understood that transdisciplinarity is not yet a theory or an institution (Jahn *et al.*, 2012), but instead simultaneously may be considered an attitude and a form of action, dealing with context specific, real world problems (Klein, 2004). Innovation in methodologies for collaborative work is central to transdisciplinarity (Mobjörk, 2010). The reported goals and outcomes of transdisciplinary practice include greater accountability through integration (Mobjörk, 2010), mutual learning and trust building (Klein, 2004), and improved disciplinary practice (Höchtl *et al.*, 2006). These are the criteria which can best be used for assessing a transdisciplinary process.

Mobjörk (2010) identifies two central approaches to transdisciplinarity in practice. First, that of consultative transdisciplinarity, where non-academic involvement is restricted to responding to research plans or results, rather than playing an active role in the research process. The second approach, participatory transdisciplinarity, emphasizes the integration of all actors in the knowledge production process. Others have differentiated the two extremes as 'symbolic' or tokenistic ('consultative') versus 'effective' ('participatory') processes (Elzinga, 2008 in Mobjörk, 2010). Transdisciplinarity processes, thus, fall on a continuum from weak to strong. These two extremes are distinguished by the level of integration and scope of the non-academic actors' roles (Mobjörk, 2010), with the latter approach more demanding and involving challenges inherent to the participatory processes (see, for example, Reed, 2008). There may be transdisciplinarity even with weak participatory processes, but when these are based on strong participation they play a key role in stakeholder integration and knowledge production, and therefore promote greater interaction. As an illustration, the analysis of the initiatives presented in this volume corresponds more closely to a consultative transdisciplinarity process, whereas the visioning process addressed in this chapter is much closer to a participatory transdisciplinary process.

Mobjörk (2010) highlights further key characteristics of transdisciplinary research projects in practice, not least the use of openness and soft-systems thinking in participant recruitment and the selection of methodologies. Similarly, soft skills are vital for the facilitation of participatory approaches in transdisciplinary research, and practitioners are also encouraged to promote reflexivity in order to respond to changes in the research context (see, for example, Mobjörk, 2010). In addition to participatory guidance, there is significant correspondence of transdisciplinarity in practice with the ideal of Communicative Action, as described by Habermas (1981). Habermas argues that social coordination relies on communication, with the aim of mutual understanding between two or more actors (Habermas, 1981; Fast, 2013). Mutual understanding is, thus, supported through the creation of an 'ideal speech situation', which ensures that all participants have the opportunity to express their views and contribute to democratic decision-making

(Brown and Goodman, 2001; Allmendinger, 2009). In transdisciplinarity, as in an ideal speech situation advocated by Habermas, all participants are equal, in other words there is no difference in the value of 'hard' and 'soft' knowledge (Lawrence and Després, 2004), and have the opportunity to share reasoned debate towards the goal of human emancipation or sustainability. The role of the researcher in this context is, therefore, to maintain an open dialogue and build close contacts with non-academic interests, as well as to participate in the research process themselves as 'stakeholders'.

However, as already highlighted in the 'traditional' participatory literature of previous decades (see Arnstein, 1969), careful attention needs to be paid to the scope of integration, with recognition that not all knowledge transfer exercises can be labelled transdisciplinary. As Arnstein's ladder demonstrates, there are 'levels' of non-participation that are contrived by some to represent genuine participation, for example, approaches which only 'inform' participants (or even manipulate them) without providing a platform for participant feedback or integration of views and ideas (Arnstein, 1969). Furthermore, it may be argued that transdisciplinarity is simply a research 'trend', promoting a rhetoric which can tend to fall low on Arnstein's ladder, given the fact that stakeholder involvement in an established research project is typically at the invitation of 'power-holding' researchers. In the era of 'post-participation', there are lessons for transdisciplinarity: to reduce ambiguity, confusion and disillusionment amongst practitioners and stakeholders, and reinforce the benefits of participation (Reed, 2008; Reed *et al.*, 2009). Further constraints arise, however, due to academic systems, limiting co-opportunities due to the research funding framework, and the perceived threat to disciplinary credibility through transdisciplinary research (disputed by Höchtl *et al.*, 2006). The movement to incorporate transdisciplinarity as a norm in professional practice is advocated in the literature, but with an acknowledged need to match aspirations held by academic publishers and funders, and an anticipation of inherent complexity (Klein, 2004).

The research reported in this volume provides support for, and rich insights into, the necessary requirements for participatory transdisciplinarity, co-construction in the research process and shared social learning to occur. These requirements emerge from an assessment of the participatory transdisciplinary process applied to a discussion on sustainability pathways, in different regions of Europe, and with different traditions for participatory processes and the science-practice dialogue.

Transdisciplinarity in practice

As explained in the introduction to this chapter, a participatory 'visioning' process supported by a transdisciplinary perspective was adopted. There were, however, underlying limitations to this approach, as the questions addressed and the methods selected did not emerge directly from practice, but were decided within the context of a pre-defined research project. Nevertheless, the research questions framing the project were defined in accordance with: (i) dialogue with practitioners and problem-solving research undertaken by the researchers; and (ii) concerns expressed by stakeholders at different governance scales in Europe, including the European Commission, which were the basis for the project calls. The method underpinning the visioning process is described below.

The identification of pathways was based on the co-construction of ideal future visions for the region (i.e. in 2030) concerning agriculture and other land-based activities. These visions correspond to scenarios, as conceptualized in the literature (Ramos, 2010). However, although the relevant literature uses the term 'scenarios', this process adopted the goal of 'visioning' in order to encourage creative and unrestricted discussion and support the creation of 'wishable futures' by the participants, rather than prediction or forecasting. These visions were not created by the research team and presented to participants for discussion and improvement; they resulted instead from co-construction. The intention behind this process was grounded in the conceptualization of system innovation and transition processes, where social learning is a process of reframing and leads to a change in perspective among stakeholders, who jointly seek a shared view of problems and directions for sustainable solutions (Kemp and Loorbach, 2006).

'Imagining the future' has been used as a way to make it easier for those involved to conceptualize the radical changes which could result from a transition process towards greater sustainability. Scenarios can be considered as tools for ordering perceptions about alternative futures (Ramos, 2010). They are usually best utilized through comparisons of different possible futures. Scenario-based approaches are particularly useful when addressing the considerable uncertainty about future trajectories in complex systems – here, uncertainty may arise from a system's complexity itself, or may be related to determining future developments (Zurek and Henrichs, 2007). They have shown to be relevant tools for improving communication amongst stakeholders, planners and decision-makers to encourage stakeholders to reflect on the future and, in this way, contribute to rural planning and sustainable governance (Tress and Tress, 2003; Carvalho-Ribeiro *et al.*, 2010; Southern *et al.*, 2011). When significant questions exist on how driving forces may play out in the longer-term future, scenarios help to explore the implications of a range of different futures. They make it possible to reduce complexity, either by looking only at part of a system, or by focusing on a concrete focal question in the scenario process.

In this research, we identify visions as the simplest form of scenarios: stories about the future which can be told either qualitatively (in words or pictures), quantitatively (as numerical estimates) or by combining both (Zurek and Henrichs, 2007). Scenarios bring together different elements of complex systems and combine them to develop images of the future. Through well-structured participatory processes, it is possible to jointly construct 'visions' which are enriched by the perspectives of different participants. This approach leads to greater awareness of the drivers in place, and also of the possible roles of different actors and institutions (Wiek and Iwaniec, 2013).

Methods

In order to ensure consistent implementation across the seven study countries, a step-by-step approach was developed (Fig. 12.1), based on participatory methodologies including individual evaluation of each step by participants. There is no pre-defined recipe for the implementation of transdisciplinary research (Brandt *et al.*, 2013). In this case, the approach was constructed for this project in particular, and adapted as it evolved, in order to achieve the best possible integration of participants and concerns, as well as to achieve the project's overarching goals. As explained in the introduction to this chapter, the selected regions were also locations of at least one of the transition case studies: Aberdeenshire

(Scotland); Plzeňský region (Czech Republic); Freiburg region (Germany); Montermor-o-Novo (Portugal); Pays de Rennes (France); Pazardjik and Plovdiv (Bulgaria); and Imathia (Greece). The goal was to convene a representative group of rural interests, including researchers, to answer two central and sequential questions:

- What is desirable for agriculture and other land-based activities for the region in 2030?
- What needs to be done to achieve this desirable future in 2030?

These two questions were formulated with different concerns: (i) to make it possible to create a distance from present conditions, construct visions which are detached from present constraints, and which could result from radical changes (in other words transitions); (ii) to think about a future which is far enough away to make a transition possible, but still close enough to be relevant for those involved; (iii) to consider agriculture as well as other activities that currently shape rural land use and the functioning of rural communities; (iv) to identify the visions and also the pathways to enable visions to be achieved; and (v) to identify the need for these pathways at different governance levels, including the regional level where discussions were undertaken. The research team defined the groups of participants to be considered according to the aims of FarmPath and the focal processes. Potential groups included farmers (those who manage agriculture at the farm level), including young farmers; those interested in and who manage other rural activities, therefore also shaping the regional context for agriculture; and those who were responsible for decision-making at the regional level. The final groups were:

- Official Interests (OI). Individuals involved in government and non-government activities related to rural issues such as; environmental organizations, farmers' organizations, established NGOs, business associations, unions, local authorities and national policy makers.
- Run the Land (RL). Individuals implementing policies though land management, therefore including farmers and land owners, hobby farmers, businesses associated with agricultural production and those responsible for protected areas.
- Young Farmers (YF). Farmers under 40 years of age who possess adequate skills to set up an agricultural holding for the first time, or are the head of the holding. This definition follows the one used in the EU rural development regulation. YF could be aggregated in the RL group but the separation was intentional to enable an assessment of whether age and accumulated experience in farming would generate different perspectives.
- Those Who Benefit from the Land (BL). This group included end users, recreational users, health-related charities, community well-being and education practitioners, social care workers, residential associations and consumer organizations.

Following this typology, individual stakeholders were contacted by the research team. During the initial contact, FarmPath objectives were presented as well as the structure of the approach (Fig. 12.1), and participants were advised on their role. This initial step was followed by a semi-structured interview to better understand each participant's background, and their individual wishes for agriculture and land-based activities in the study region for

2030. This step was designed to build up trust between the research team and participants, so that the latter could feel that their voices were heard and well integrated into the process from the outset.

The second step was a focus group discussion with individuals from the same group, where the main concerns and discourse were expected to be similar. This contributed to a smoother introduction to the methodology for participants and researchers. The aim was to produce visions for agriculture and other land-based activities for each region in 2030, and to identify the transition pathways necessary in order to achieve those visions. As such, the exercise was based primarily on a normative approach to the future, questioning 'what should happen', and providing a perspective on scenario building appropriate to strategic assessments (Ramos, 2010). The exercise was based on systems thinking and conceptual modelling, using drawing of circles and arrows with text content (Guimarães *et al.*, 2013), to allow a structured discussion and ensure that all dimensions of the question were covered. The conceptualization of this step was designed to be inclusive, ensuring that all participants could express themselves and felt heard (Wiek and Iwaniec, 2013).

The step-by-step approach culminated in a final workshop, including all participants (all stakeholders from the four groups, the members of the NSPG and the researchers). The aim was the co-construction of pathways to achieve previously defined visions. After the workshop, the pathways created were analysed by the research team using standard qualitative analysis techniques and further supplemented by discussion with the NSPG.

It is important to note that the focus groups and the workshops were supported by professional facilitators. This made it possible for researchers to fully engage in the discussions but more importantly, it has ensured that all those involved were secure about their role and the expectations surrounding their contributions.

The visions and pathways identified

The visions

More than 50 visions were gathered across the seven European regions. Even considering the large differentiation of the regions, there were many similarities in the way these visions could be grouped using their central foci. There were also fundamental differences which can be partly explained by the particularities of the regional contexts (see Table 12.1).

One group of visions (including eight of the visions identified) can be summarized as the intensification of production, neo-productivism, farming competitiveness and profitability. Within these visions, environmental constraints were expressed, but the focus was on farming production and productivism as a key strategy. Another group of visions (three of the visions identified) related to farming, but were centred on the environmental or conservation agenda, with the quality of the landscape and of the environment or natural resources as an expression of the desired outcomes. Finally, a third group of visions (comprising eight visions) focused on rural communities, a lively countryside, networks and close connections between the urban and the rural, strongly emphasizing rural values and lifestyles. Many visions were primarily centred on one of these three dimensions, but also included elements of others.

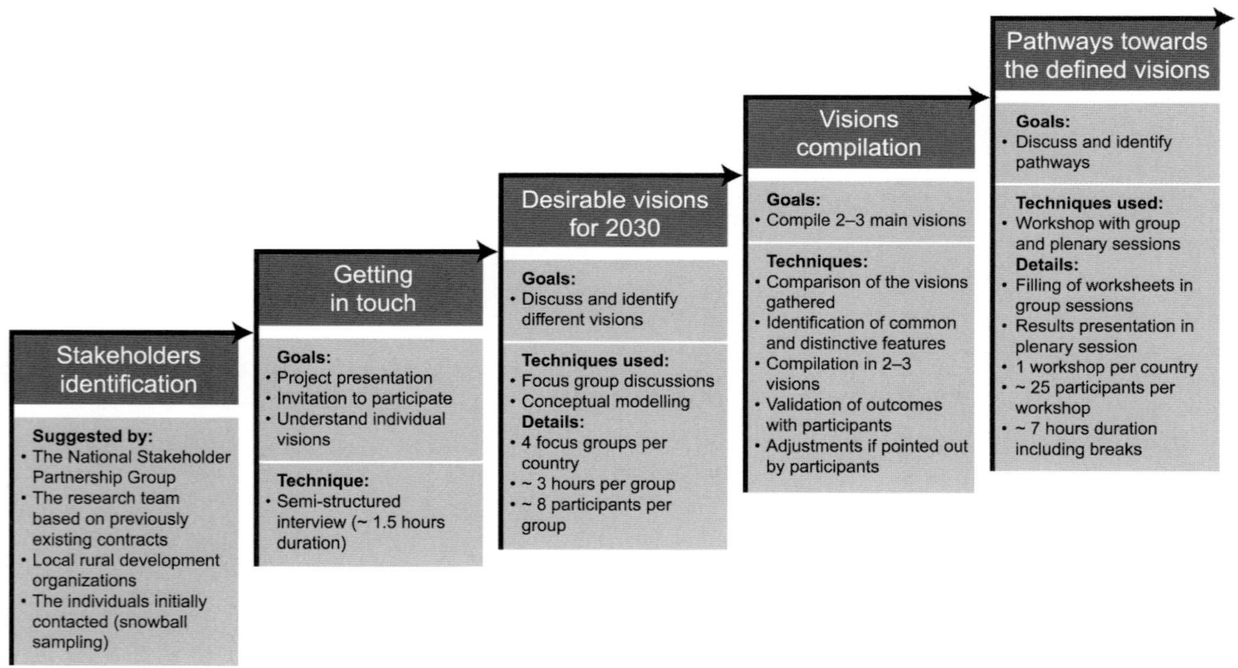

Fig. 12.1. Step by step implementation approach (Source: authors).

Table 12.1. The visions (Source: authors).

<table>
<tr><th colspan="5">Vision typologies and key points of each vision created</th></tr>
<tr><th>CASE STUDIES</th><th colspan="2">Intensification of production, neo-productivism, farming competitiveness and profitability</th><th>Farming centred on a conservation agenda, landscape quality as a desired outcome</th><th>Lively countryside with vibrant rural communities, and strong rural-urban networks. Strong reinforcement of rural values and lifestyle</th></tr>
<tr><td>Aberdeenshire (Scotland)</td><td colspan="2">Farm resilience, profitability, public payments for non-market goods, respecting environmental standards.</td><td>Food security, re-connecting people with land, diversified farms, environmentally friendly, reduced fossil fuels.</td><td>Connected communities, urban-rural networks, innovative housing, infrastructure.</td></tr>
<tr><td>Plzeňský region (Czech Republic)</td><td colspan="2">Food production, economic viability, protected domestic markets, higher status of farming, food quality.</td><td>Prosperous countrysides, prominent small scale production, rural tourism, cultural landscape.</td><td>Strong social dimension, cooperation between farmers, direct support of rural living for small farms and young farmers.</td></tr>
<tr><td>Freiburg region (Germany)</td><td colspan="2">‘Regional competitiveness’, ‘environment-friendly management’, energy conservation and production.</td><td>Cultural landscape, society values agriculture, long-term policies, economic viability.</td><td>People-centred agriculture, ‘sustainability’ and ‘individual self-responsibility’. Closer relationship between society and agriculture.</td></tr>
<tr><td>Montermor-o-Novo (Portugal)</td><td colspan="3">Intensification of production, economic viability. Safeguard the Montado system, better technology, new rural identity.</td><td>Preserve Montado system, regional trademark, new mind-set and identity, cooperation, strategic planning, training.</td></tr>
<tr><td>Pays de Rennes (France)</td><td colspan="2">Large competitive farms, farmer-managed processing, no urban sprawl.</td><td colspan="2">Small farms, diversification of production and rural activities, alternative market channels, cooperative organization, micro-industry, energy production.</td></tr>
<tr><td>Pazardjik and Plovdiv (Bulgaria)</td><td>Efficiency, low environmental limits, new tech.</td><td>Specialization, intensification, exports.</td><td></td><td>Economic efficiency, cooperation and interaction social cohesion, local brand, direct marketing. Better quality of rural life.</td></tr>
<tr><td>Imathia (Greece)</td><td colspan="2">Modernization and specialization of farming. Establish quality brand name.</td><td>Biodiversity, specialization, training, collaboration, quality products.</td><td>Integrated rural development, environmental spatial planning, natural resource protection cultural heritage management; better quality of life.</td></tr>
</table>

The community was seen as a source of strength and differentiation in the countryside, and the social dimension appeared to be the driver for at least one vision in the seven regions considered; it is therefore the most commonly shared ideal for the future. Furthermore, increased production and economic competitiveness appeared as a goal but was not given equal importance in every region. In the Czech Republic case, two visions emphasized social dimensions: 'Agriculture for the countryside', stressed cooperation and networking amongst farmers, and another 'Lively Countryside', emphasized multifunctionality and a diversified rural community. Both were centred on the rural community. The Bulgarian case was an exception, as all three visions aimed for intensification, modernization and specialization. One vision included elements of quality of life in rural areas and in all three, neo-productivism was central. This particular focus in Bulgaria may have several explanations, including the relatively low level of agricultural modernization in the country, or may reflect the profile of those involved in the process. In Portugal, the two visions created can be included in two of the groups described above, although both included the Montado as a central condition for future sustainability. Montado is the extensive silvo-pastoral land use system characteristic of southern Portugal and in both visions its maintenance was considered as fundamental. Broadly, the different characteristics of the regions do not seem to be reflected in clear differences in the type of visions formulated. Therefore, despite the differentiation of regions across Europe, not least their variation in the dominance of intensive or extensive agricultural systems, common visions emerged, although with multiple potential outcomes due to regional specificities.

The pathways

The pathways necessary in order to achieve these visions reveal a much wider diversity of concerns. The discussion in the final workshop, in each of the regions, resulted in a large and multiple list of pathways, which can be summarized:

- Maintenance or re-emergence of farming activities;
- Innovation in farming;
- New concepts of farming, farmers and rural areas; and
- Overall policy and institutional change.

The maintenance or re-emergence of farming activities means that regardless of the farming system, there are certain current features considered to be essential to maintain, or re-activate, the social and economic role of agriculture. It was felt that this could be achieved through the development and maintenance of farming infrastructure and services, the economic viability of farming activities, well-planned land and farming succession, and closer interconnections between farming, policy and research.

Innovation in farming was considered achievable through innovative mind-sets and practices, and through the use of new techniques and technologies, practices and network connections; all of which were considered necessary for the future sustainability of agriculture and for other land-based activities.

'New concepts' referred to the need to acknowledge the shift away from production as the sole driver of land use and rural dynamics towards a complex interplay of other drivers, such as countryside consumption, or landscape and nature conservation. This pathway

focuses on the multifunctional nature of the ideal transition and the need for multifunctionality to be acknowledged by public policies, as well as in the recognition of the range of actors involved in decision-making and management. The conditions required for these new concepts to spread are 'reshaped relations' between farming and the wider public, based on the attractiveness of rural areas, the trend for 'going local' (for buying locally produced goods), and for farming to be reintegrated into the local community. Multifunctionality was also seen as a central concept for farming and rural areas, with integrated actors and strategies.

Policy and institutional arrangements were those conditions that must be established at the macro level which frame the activities to be developed in rural areas. These arrangements correspond to the different sectors and strategies transferred into activities and legislation at different scales. For example, targeted rural investment and changing farmer mind-sets regarding the involvement of local communities were needed in order to gain Scottish Rural Development Programme (SRDP) funding. It was felt that these arrangements are best achieved through coherent policy-making, regulation, funding, institutions, and integrating understanding of, and prioritizing, global policy issues.

Although the 'desirable' futures showed remarkable similarities across the study regions, the wide range of actions suggested indicates the importance of contextualized intervention and action, adapted to the characteristics and needs of each region. In such a diversified rural Europe, this outcome could be anticipated, yet it is still striking, nonetheless. The margin for flexible regulations and tailor-made solutions is increasingly small, constrained as it is by international agreements and European policy making, together with the growing globalization of markets and models. The results show that the opposite is considered necessary; that in fact, specific solutions and combined actions are required for sustainable pathways.

The way it worked: transdisciplinarity in participatory approaches

The utilization of the same step-by-step methodology across seven regions makes it possible to assess the key points of success and failure of the approach and relate them to the context of application.

According to the evaluation by participants at each step of the process, the reasons behind participants' interest in the project vary between tangible (e.g. to facilitate changes in policies, management actions) and intangible aspects (e.g. to meet other people, work with scientists). Even participants with experience of participatory projects valued the networking potential of this project. Conversely, several participants, including the researchers, expressed worries about achieving tangible results from the project.

This point leads to a relevant question regarding the motivations for transdisciplinary work, as well as the perspectives of those who promote it. Despite the fact that tangible results are the ultimate goal (e.g. policy development) such outputs do not only depend on the transdisciplinary process. This misunderstanding can contribute to problems which are well documented in the stakeholder participation literature, such as discredit and fatigue (Reed, 2008). As shown by several different authors, a transdisciplinary process requires the commitment of all participants, and hence it must be coupled with sharing responsibilities and empowerment of all participants (Brandt *et al.*, 2013).

In the FarmPath process, the percentage of those who agreed to participate in the first step and continued until the final stage varied from 44% to 63% in different countries. If participant motivation was not high during the recruitment stage it generally resulted in non-attendance at later stages (particularly the final workshop) despite agreement to participate. Careful selection, reinforced by researchers' assessment of each individual, was required at the recruitment stage in order to ensure that the majority of participants were highly motivated. Furthermore, careful structuring of the process to ensure that participants felt valued and got something meaningful out of the process was crucial to secure their continued engagement. Perhaps more interesting is the fact that in regions where participants knew the research team previously, participation was more consistent. This happened in the Czech, French and Portuguese cases, where many participants had previously been involved in projects lead by the research team. This seems to indicate that social capital can be accumulated between projects, and exceeds the time frame of a project; a research group engaging in such an approach is developing a bond with the stakeholder community, which can be reinforced at each project when results are enriching for all parties, as well as broken if partners are disappointed.

Within the focus-group discussions in the different regions, 'Official Interests' (OI) was the group in which the visioning exercise was most easily achieved. This was considered to be because these individuals engage in strategic thinking exercises as part of their professional life; they were therefore familiar with thinking in terms of visioning. Their discourse was already well-structured. This was not the case in the remaining focus groups, many of which produced positive comments on the novelty of approach, as many had not previously been engaged in participatory processes. On the other hand, the group 'Those who Benefit from the Land' (BL) expressed greater difficulties in creating regionally based visions and acknowledging the role of farming in the construction of the physical landscape, probably due to their weak connection with the sector. Nonetheless, during the process some questions were clarified and the discourses of some participants changed. These results reveal the suitability of a step-by-step approach in a transdisciplinary exercise, to allow participants to become familiar with the process. Aiming for a co-construction process involving actors from different spheres of society does not only imply difference in knowledge types, but also in the maturity of discourses. Hence opportunities to better structure individual and group discourses are necessary in order to promote subsequently balanced dialogues.

The construction of visions, or the design of desirable futures not rooted in present situational constraints, was viewed as the most challenging task by participants. It confirmed that innovative thinking is not easily achieved and the support of different specific strategic methods is important if the visioning exercises are to be successful. Nevertheless, the systemic approach to vision building, with a conceptualization followed by the drawing of circles and arrows as described in the Methods section, was considered by some as a schematic exercise which decreased the level of in-depth discussion. In addition, the need to include more intervention of researchers in the vision-building step was mentioned. During this step, the researchers in some countries were solely observers (being involved in the discussion only in the later phase) but in the countries where the researchers also engaged in the discussion by adding questions and providing personal perspectives, this constraint was not mentioned by participants.

Regarding the 'visions compilation' step, despite the fact that most participants validated the results achieved, several indicated that the way it was organized decreased the motivation of participants to continue. As demonstrated through the visioning process in many countries, passive interaction tools, such as emails and written documents, may be considered to have minor co-construction capacity.

The final workshop was designed specifically to facilitate a participatory transdisciplinary exercise, as it included interaction between members of all groups, NSPG members and researchers. Several researchers expressed difficulty in joining the discussion as participants, and the remaining participants expressed some difficulty in considering the researchers as participants, especially when they were in disagreement. Researchers were seen as those 'who know', 'who have the knowledge', and therefore were hard to contradict. Since most formal transdisciplinary projects are promoted by science partners, this suggests the need for better preparation on the part of researchers so that the power of academic knowledge is in balance with others, and also that the role of the scientist in the process is understood. In addition, it is important to engage professional facilitators in participatory methods, so that researchers can fully engage in the process.

The evaluation by participants in all countries concluded that the transdisciplinary process created engagement and in-depth discussions. However, participants began the process with highly variable levels of experience of transdisciplinarity. Where there was previous experience of involvement, particularly in European projects, participants showed some fatigue in relation to this kind of project. In others, the novelty of the process, and in particular the European dimension, has raised great interest. As previously described by authors such as Brandt *et al.* (2013), and as found in this research, willingness to participate and consistency of participation rely on a common agenda defined with the involvement of all participants. However, such common agendas require the novel design of research projects, with the use of different strategies and methods established together at the outset, which may not be possible in existing research frameworks.

Creating a parallel to the multi-level perspective (MLP), transdisciplinarity can be seen as a niche in an anchoring process within the established regime of science and research. Multiple linkages are established between different niches, sub-regimes and the regime. This also includes a pattern of interactions between science and society that is not yet established and may, in some cases, hinder the outcomes of transdisciplinary efforts.

How can transdisciplinarity support transitions to sustainability in agriculture?

The transdisciplinary process developed within FarmPath has been conducted over a relatively short timescale, therefore an overarching assessment of its impacts is not possible to undertake at the time of writing. However, it has empowered local participants, leading to a greater awareness of their possible roles in a potential transition. In addition, it has led to the acknowledgment by practitioners that links to science can be useful and inspiring, and are easily established. Furthermore, and what may be more interesting, this experience has revealed to researchers how interaction with practitioners can be achieved, and how fruitful it can prove to be. It is not feasible to assess how researchers have changed, but from the evidence gathered and the reports delivered, it can be concluded that a more

positive attitude towards co-construction processes has been achieved. Even with a cursory assessment of its direct value, it is possible to say that the transdisciplinary process has enhanced the science-society dialogue and thus contributed to transcending the science-society divide (Neef and Neubert, 2011; Darnhofer, this volume).

In most regions (Montemor-o-Novo in Portugal, Pays de Rennes in France, Freiburg in Germany, Imathia in Greece, Plzeňský in Czech Republic and Aberdeenshire in Scotland), participants of the final workshop expressed their desire to continue to be involved in discussing pathways for the future sustainability of agriculture in their region, emphasizing the need to maintain an open dialogue and for co-construction coordinated by the researchers. The role of researchers is viewed not only as positive, but also as a condition enabling the process to continue. This reveals an expectation concerning the active role of researchers and indicates that outcomes from a transdisciplinary dialogue can go much further than the achievement of goals within a single project or collection of information for scientific purposes.

In some cases, ability to participate and the ease with which participants understood the research questions and project discourse, were related to existing and long-term professional relationships between the research team and many of the participants involved. Trust capital 'build–up' over the long term should be acknowledged, and can be considered to be long term transdisciplinarity. The effects of each transdisciplinary participatory process, such as those developed through FarmPath, need to be understood in the context of the long term interaction between science and practice.

The role of transdisciplinarity in transitions cannot be understood without considering the role of science, the scientific spheres and the scientific actors, in transition processes. The MLP describes how the rules within one regime orient and coordinate the activities of various sub-regimes, such as policy, science or socio-cultural sub-regimes (Geels, 2011). While sub-regimes share core guiding values, they also have their own specific dynamics and rules. Thus, it is expected that there will be differences between, for example, policy makers and researchers within the agricultural regime. A specific sub-regime may, therefore, play an important role in building tensions within the regime and supporting or hindering an emerging transition. Furthermore, researchers are also part of other regimes, including the science regime, and may be under the influence of other sets of rules and guiding values. They may, therefore, act outside or as hybrid actors in a transition process in agriculture, again creating the tensions noted above within the regime, which may be needed to open up possibilities for niches to anchor.

The previous chapters have shown how different innovative niches in many different regions of Europe have managed to link to the regime, and what supports or hinders their take-off and anchoring. Considering these cases, the MLP shows us that transitions in agriculture are complex and need to address many more sectors and actors than those directly related to agriculture. In order for a connection from a niche to sub-regimes or regimes to occur, we have seen how different actors need to be mobilized at the niche level, at the regime level, and also in hybrid positions. These hybrid positions are often those adopted by researchers, outside the specific practical issue but in close interaction with it, and with relevant or significant knowledge of its 'contours' or specificities. In addition we have seen, in many of the niches studied, that local actors look for supporters to defend their interests; who can connect them to other spheres of knowledge or influence; and who can help the anchoring phase. Researchers are often chosen to act as supporters. In other

cases, researchers have been active social players, in starting the niche or creating the dynamics required for the niche to take off. Researcher involvement can result in higher social capital amongst those involved which, in turn, can also be the driver for more hybrid connections and interactions amongst niches, sub-regimes and regimes. Researchers' networking capacity, conceptual background, knowledge base and position as relative outsiders ensures that researchers can be key actors in many phases of the transition process. What the visions have also shown is that transition is considered to be possible only when anchoring of the niches can take place in different sub-regimes or regimes, and these can interact – increasing the complexity of the process and the need for hybrid actors, as researchers may be, to play an active role.

Conclusion

As advocated by Marsden *et al.* (2010; Marsden, 2013), the participatory transdisciplinary process has attempted to identify visions and pathways for the sustainability of agriculture at the regional scale, and also to support more reflexive and adaptive governance in the regions concerned. Nevertheless, the approach has not resulted from the joint understanding of a problem, where both researchers and practitioners have, together, defined the need for the process, as is the ideal for co-construction (Darnhofer *et al.*, 2012). Therefore, the scope of this process was limited from its initiation. Nonetheless, it can be concluded from the findings of the implemented process that social learning has occurred in each case study region, involving multiple actors from different spheres. This learning is more clearly identified in some regions than in others, but the co-construction of visions and pathways, resulting from the process as a whole, inherently contributes to social learning (Grin *et al.*, 2010). Results have also shown that this process has led to a change in perspective amongst stakeholders, including the researchers, through joint efforts to find a collective perception of the problem, and directions for sustainable solutions (Kemp and Loorbach, 2006). Consequently, the transdisciplinary process has contributed to changes and possibly, in some regions, has supported a transition pathway. Only future assessments will be able to confirm this statement.

Furthermore, this social learning has only been possible due to the use of suitable and tailor-made tools. It became obvious, as Brandt *et al.* (2013) also found, that a well-structured and facilitated process, where the leading role is clearly defined, is a critical factor in the progression of the shared construction of knowledge. This critical factor also encompasses the attitudes and behaviours of the research team. Only when these are open to the science-practice dialogue can such crucial dialogue like this take place and be successful.

References

Allmendinger, P. (2009) *Planning Theory (Planning, Environment, Cities)*, 2nd edn. Palgrave MacMillan, Basingstoke, UK.

Arnstein, S.R. (1969) A ladder of citizen participation. *Journal of the American Institute of Planners* 35, 216–224.

Brandt, P., Ernst, A., Gralla, F., Luederitz, C., Lang, D., Newig, J., Reinert, F., Abson, D.J. and Wehrden, H. (2013) A review of transdisciplinary research in sustainability science. *Ecological Economics* 92, 1–15.

Brown, R.H. and Goodman, D. (2001) Jürgen Habermas' theory of communicative action: An incomplete project. In: Ritzer, G. and Smart, B. (eds) (2003) *Handbook of Social Theory*. SAGE Publications, London, UK, pp. 201–216.

Carew, A. and Wickson, F. (2010) The TD wheel: A heuristic to shape, support and evaluate transdisciplinarity research. *Futures* 42, 1146–1155.

Carvalho-Ribeiro, S.M., Lovett, A. and O'Riordan, T. (2010) Multifunctional forest management in northern Portugal: Moving from scenarios to governance for sustainable development. *Land Use Policy* 27, 1111–1122.

Darnhofer, I. (2015) Socio-technical transitions in farming: Key concepts. In: Sutherland, L-A., Darnhofer, I., Wilson, G.A. and Zagata, L. (eds) *Transition Pathways towards Sustainability in Agriculture: Case Studies from Europe*. CABI, Wallingford, UK, pp. 17–32.

Darnhofer, I., Gibbon, D. and Dedieu, B. (2012) Farming systems research: An approach to inquiry. In: *Farming Systems Research into the 21st Century: The New Dynamic*. Springer, Dordrecht, the Netherlands, pp. 3–31.

Darrot, C., Diaz, M., Tsakalou, E. and Zagata L. (2015) 'The missing actor': Alternative agri-food networks and the resistance of key regime actors. In: Sutherland, L-A., Darnhofer, I., Wilson, G.A. and Zagata, L. (eds) *Transition Pathways towards Sustainability in Agriculture: Case Studies from Europe*. CABI, Wallingford, UK, pp. 143–156.

Fast, S. (2013) A Habermasian analysis of local renewable energy deliberations. *Journal of Rural Studies* 30, 86–98.

Geels, F.W. (2011) The multi-level perspective on sustainability transitions: Responses to seven criticisms. *Environmental Innovation and Societal Transitions* 1, 24–40.

Grin, J., Rotmans, J. and Schot, J. (eds) (2010) *Transitions to Sustainable Development. New Directions in the Study of Long Term Transformative Change*. Routledge, London, UK.

Guimarães, M.H., Ballé-Béganton, J., Bailly, D., Newton, A., Boski, T. and Dentinho, T. (2013) Transdisciplinary conceptual modeling of a social-ecological system - A case study application in Terceira Island, Azores. *Ecosystem Services Journal* 3, 22–31.

Habermas, J. (1981) *The Theory of Communicative Action Volume 1: Reason and the Rationalization of Society*. Heinemann, London, UK.

Healey, P. (2006) *Collaborative Planning: Shaping Places in Fragmented Societies,* 2nd edn. Palgrave MacMillan, Basingstoke, UK.

Höchtl, F., Lehringer, S. and Konald, W. (2006) Pure theory or useful tool? Experiences with transdisciplinarity in the Piedmont Alps. *Environmental Science and Policy* 9, 322–329.

Jahn, T., Bergmann, M. and Keil, F. (2012) Transdisciplinarity: Between mainstreaming and marginalization. *Ecological Economics* 79, 1–10.

Karanikolas, P., Vlahos, G. and Sutherland, L-A. (2015) Utilizing the multi-level perspective in empirical field research: Methodological considerations. In: Sutherland, L-A., Darnhofer, I., Wilson, G.A. and Zagata, L. (eds) *Transition Pathways towards Sustainability in Agriculture: Case Studies from Europe*. CABI, Wallingford, UK, pp. 51–66.

Kemp, R. and Loorbach, D. (2006) Transition Management: A reflexive governance approach. In: Voβ, J.P., Bauknecht, D. and Kemp, R. (eds) *Reflexive Governance for Sustainable Development*. Edward Elgar Publishing, Cheltenham, UK, pp. 103–130.

Klein, J.T. (2004) Prospects for transdisciplinarity. *Futures* 36, 515–526.

Lang, D.J., Wiek, A., Bergmann, M., Stauffacher, M., Martens, P., Moll, P., Swilling, M. and Thomas, C.J. (2012) Transdisciplinary research in sustainability science: Practice, principles, and challenges. *Sustainability Science* 7, 24–43.

Lawrence, R.J. and Després, C. (2004) Introduction: Futures of transdisciplinarity. *Futures* 36, 397–405.

Marsden, T. (2013) From post-productivism to reflexive governance: Contested transitions in securing more sustainable food futures. *Journal of Rural Studies* 29, 123–134.

Marsden, T., Lee, R., Flynn, A. and Thankappan, S. (2010) *The New Regulation and Governance of Food: Beyond the Food Crisis?* Routledge, London, UK.

Mobjörk, M. (2010) Consulting versus participatory transdisciplinarity: A refined classification of transdisciplinary research. *Futures* 42, 866–873.

Neef, A. and Neubert, D. (2011) Stakeholder participation in agricultural research projects: A conceptual framework for reflection and decision-making. *Agriculture and Human Values* 28, 179–194.

Pohl, C. and Hirsch Hadorn, G. (2007) *Principles for Designing Transdisciplinary Research.* Swiss Academies of Arts and Sciences, Oekom, Munich, Germany.

Ramos, I.L. (2010) Exploratory landscape scenarios' in the formulation of landscape quality objectives. *Futures* 42, 682–692.

Reed, M.S. (2008) Stakeholder participation for environmental management: A literature review. *Biological Conservation* 141, 2417–2431.

Reed, M.S., Graves, A., Dandy, N., Posthumus, H., Hubacek, K., Morris, J., Prell, C., Quinn, C.H. and Stringer, L.C. (2009) Who's in and why? A typology of stakeholder analysis methods for natural resource management. *Journal of Environmental Management* 90, 1933–1949.

Röling, N. and Wagemakers, M.A. (eds) (1998) *Facilitating Sustainable Agriculture. Participatory Learning and Adaptative Management in Times of Environmental Uncertainty.* Cambridge University Press, Cambridge, UK.

Schiller, S., Gonzalez, C. and Flanigan, S. (2015). More than just a factor in transition processes? The role of collaboration in agriculture. In: Sutherland, L-A., Darnhofer, I., Wilson, G.A. and Zagata, L. (eds) *Transition Pathways towards Sustainability in Agriculture: Case Studies from Europe.* CABI, Wallingford, UK, pp. 83–96.

Southern, A., Lovett, A., O'Riordan, T. and Watkinson, A. (2011) Sustainable landscape governance: Lessons from a catchment based study in whole landscape design. *Landscape and Urban Planning* 101, 179–189.

Spangenberg, J.H. (2011) Sustainability science: A review, an analysis and some empirical lessons. *Environmental Conservation* 38, 275–287.

Tress, B. and Tress, G. (2003) Scenario visualisation for participatory landscape planning - a study from Denmark. *Landscape and Urban Planning* 64, 161–178.

Wiek, A. and Iwaniec, D. (2013) Quality criteria for visions and visioning in sustainability science. *Sustainability Science*. DOI: 10.1007/s11625-013-0208-6.

Zurek, M.B. and Henrichs, T. (2007) Linking scenarios across geographical scales in international environmental assessments. *Technological Forecasting & Social Change* 74, 1282–1295.

Chapter 13

Conceptual insights derived from case studies on ‘emerging transitions’ in farming

I. Darnhofer[1], L-A. Sutherland[2], T. Pinto-Correia[3]

[1] *University of Natural Resources and Life Sciences, Vienna (ika.darnhofer@boku.ac.at);*
[2] *James Hutton Institute, Aberdeen;* [3]*University of Évora*

Introduction

The multi-level perspective (MLP) has been applied to agri-food studies in the past, however, these studies have tended to focus either on large-scale (national) and often historical transitions (e.g. Grin, 2010), or on very specific system innovations initiated by technical innovations (e.g. Elzen *et al.*, 2012). The former focus on long-term processes and, thus, allow an assessment of whether or not a broader transition has occurred, or at least is well under way. The latter focus on processes that started recently, and whose future development is still uncertain so that it may be too early to speak of a ‘transition in the making’. The ambition of the case studies presented in this book was to provide novel insights on the ‘middle-ground’: niches that have matured and that have started engaging with the regime to initiate the take-off phase of a transition. The coverage of very diverse cases, from different regions of Europe and relating to different combinations of regimes, was designed to acquire a broad understanding of emerging transitions occurring in relation to farming and rural areas in Europe.

In this chapter, we will briefly review the specificity of farming and associated implications for understanding transitions. We will then review the lessons learned, especially regarding the definitions of niche and regime in the context of transitions, and in relation to niche-regime interactions. We will also briefly reflect on the role of research – as part of the regime – in emerging transitions. We conclude by assessing the challenges of studying transitions in farming, and the challenges posed by the need to take into account the diversity within farming.

The specific features of farming

There are several features that distinguish the agricultural sector from the industrial or service sectors which need to be taken into account when studying transitions. We will briefly review a few of these features, which include diversity in farming, its spatial nature, its multifunctionality and its public good character, all of which contribute to the high level of policy involvement in the sector.

Farming is fundamentally a land-based activity, and is, therefore, heavily shaped by the local agri-ecosystem, topography and climate, as well as by traditions, economic structures and social norms that have co-evolved in this natural environment. As natural conditions and cultural traditions vary, both within a region and between regions, farming structures, practices and values are very diverse (see Slee and Pinto-Correia, this volume). Indeed, on the farm various activities can be integrated in different ways to suit local conditions and farmer preferences. As a result, within a given region farms vary in size, activity and market-orientation. Some farmers will focus on reducing production costs to increase their competitiveness on international commodity markets, whilst others diversify their crops or engage in processing to provide a broad and attractive range of products for sale at farmers' markets. Yet others may opt for 're-grounding' by diversifying the economic activities based on the farm. Between regions, diversity is also pronounced; ranging from small, 0.5 ha semi-subsistence farms in Bulgaria, to 250 ha mixed crop-livestock farms in southern Portugal, to 10,000 ha estates in the UK. Diversity in farming is not limited to production methods, farm structures, or the influence of terrain and climate; it is also influenced by the types of markets that farmers serve, be they long or short food chains, energy markets, or the services that they offer (e.g. tourism and recreation). Given this diversity, a transition is not likely to lead to a uniform set of practices but rather to a different mix with different emphases, and differences in the linkages between elements of the farm system. This might make it challenging to pinpoint a clear transition from a set of practices 'A', to a new – and radically different – set of practices 'B'.

Marsden (2013) has also pointed out the inherently spatial nature of farming systems. As such, the biophysical conditions are as decisive as is the location of a region (especially whether it is peripheral or close to a large urban area). Both can play an important role in the types of transitions that are more likely to 'take-off'. For example, the case studies on 'countryside consumption' (Pinto-Correia *et al.*a, this volume) show that a transition from agricultural land being used primarily for food production towards agricultural land being used primarily as a living space and for recreation, is more likely to occur in regions with an attractive physical landscape close to an urban centre, not least as the latter will offer jobs and, thus, income opportunities. Similarly, urban centres play an important role in the cases linked to alternative agri-food networks (Darrot *et al.*, this volume): the proximity to large cities and the demand by urban consumers has enabled farmers to co-construct direct marketing initiatives with them. In some cases, the spatial nature of agriculture might seem less relevant; for example in the production of renewable energy, as the electricity generated can be fed into the grid and, thus, transported over large distances (Sutherland *et al.*b, this volume). However, even in that case the low quality of grid connections in remote areas tend to privilege more centrally located farms, whilst peri-urban developments are more likely to be subject to public protest.

Another feature is linked to the function of farming. Indeed, farming is often perceived as having primarily one societal function: food production, and, therefore, as being one clearly defined sector. Within the MLP, this sector can be understood as a socio-technical regime, with its constituting sub-regimes (see Geels, 2011), for example: agricultural policy, agricultural research, the agri-food industry, food production and processing technology, market and consumer preferences. However, this sectoral (or food-chain) understanding of agriculture has been the object of discourse in Europe since the 1990s, when ideas surrounding the multifunctionality of farming came to the fore (Marsden and Sonnino, 2008; Renting *et al.*, 2008). Although the specific aspects of multifunctionality are debated, there is broad acceptance of the general principles. These state that whilst the function of farms is to produce food and other goods (e.g. fibres, energy), they also provide other (non-market) functions such as protection of natural resources (soil, water, biodiversity), maintenance of forests, biotopes and other valued elements of landscapes; as well as contribute to the cultural heritage of rural areas (including traditional and speciality foods). The understanding of farming as multifunctional, therefore, shifts the perspective from a sectoral to a territorial approach, where the manifold interdependencies of rural areas and farming are emphasized.

This multifunctionality was highlighted in a number of the case studies presented in this book, where it became clear that the niches studied resulted from interactions between the agri-food regime and other regimes, such as recreation, energy or environmental protection. As Holtz *et al.* (2008) have pointed out, niches may emerge through novel interactions between regimes. Given its multifunctionality, it would seem that in farming, potential transitions are likely to involve niches that emerge from such novel interactions. These interactions are mostly present at the level of the farm, from which very diverse value chains may emerge, linked to aspects such as recreation, energy production (through windmills, wood or biogas) or food production (as commodity for the food industry, or as traditional food sold directly to consumers).

The fact that farms contribute to several crucial functions in society, that farmers are stewards of over half of Europe's territory and that they produce many public goods, all contribute to a high level of policy involvement. Indeed, agriculture was one of the first sectors where policy was made at EU-level, not least driven by the importance of securing food supply after the Second World War. While initially, the aim of the Common Agricultural Policy (CAP) was food security, it later focused on the competitiveness of farms, and – since the Agenda 2000 – has also aimed to ensure environmental sustainability. Since its beginning, the CAP was also a 'farm policy', aimed at contributing towards a fair standard of living for farmers (European Union, 2012). Indeed, the European vision for rural areas is that they are integrated into the economy through their resources and local initiatives, incorporating economic, social and cultural dimensions. The aim is to ensure that rural areas are a place with dynamic communities, and to keep 'lights on in the windows', even in remote rural areas. The second pillar of the CAP has promoted the diversification of farms, thereby encouraging farmers to go 'beyond' agriculture and addressing emerging societal expectations, whilst opening up new sources of income.

These rural development policies have opened up windows of opportunity which have allowed 'dormant niches' to become more visible, and new niches to develop. Thus, the agricultural side can be 'deepened' through engaging in value-added activities, such as short food chains or organic farming; the rural side of the farm might be 'broadened' by creating new income flows to the farm enterprise, for example through new on-farm activities such as agri-tourism, or care farming; the farm can be 're-grounded' in a new pattern of resource use, for example through pluri-activity (van der Ploeg and Roep, 2003). As such, many of the case studies included in this book are an illustration of niches co-dynamically developing at the interface between 'bottom-up' and 'top-down' processes. Indeed, while niches are often seen as struggling to take-off in the face of a reluctant (and often 'locked-in') agricultural regime, within farming it seems that the policy sub-regime is actively engaged in promoting change – even if 'change' should not be equated with 'transition', and even if these forces are not necessarily dominant.

The close link between farming and its (natural and social) context, its spatial dependence, the diverse functions it fulfils and the diverse societal expectations it faces, means that the agricultural regime is neither homogeneous nor monolithic (see Smith *et al.*, 2005). Rather, as the case studies presented in this book emphasize, the regime (that is the policy paradigms, visions, social expectations and norms) are 'semi-coherent' and characterized by internal tensions, disagreement and conflicts of interests (Geels, 2011). Thus, whilst the agricultural regime is focused on promoting 'modernized agriculture' (or might want to appear that way), at the same time it offers some protection for niches, such as by providing funds and regulatory support for rural development activities like alternative agri-food networks (AAFNs). In farming, transitions are likely to be characterized by diversity, and to result from push-and-pull efforts by niche actors in cooperation with regime actors (of the agricultural or other regimes).

Specifying the MLP for the farming context

Which niche?

Within the MLP, the niche is the place where entrepreneurs work on radical innovations that deviate from established norms, processes or practices, usually based on different beliefs. As a result of the adjustment of expectations, enrolment of new actors, expansion of the resource base and learning processes, networks become larger, especially through the participation of powerful actors who convey legitimacy and supply additional resources (Geels, 2011). Niche actors are usually understood as being driven by the hope that their novelty might eventually be used in the regime or even replace it (Geels, 2011). Upon closer analysis, a number of case studies reported in this book deviated in some way from this typical characterization of a 'niche': some did not display a clear transformative ambition; some did not start from the 'bottom up'; in others, the actors were not in a coordinated network.

Interestingly, in a number of the cases included in this book, the ambition of the initiative studied was not to transform the regime. Rather, the aim was for topical and local change which would allow actors to pursue their project, without ambition to engage with, or question, the dominant practices (Fig. 13.1). For example, the network engaged in establishing farmers' markets in Pilsen (see Darrot *et al.*, this volume) was not particularly interested in questioning dominant food purchasing practices or dominant actors (e.g. supermarkets). Its aim was primarily to build a space for its own project, and propose an additional option for purchasing fresh food (especially in summer), rather than to initiate a wider transition. In some cases, alternative agri-food networks might even put particular emphasis on retaining their autonomous status (ensuring that they have little or no ties to mainstream actors, thus purposively not engaging with the agricultural regime). The alternative marketing channels are thus an example of initiatives that do not aim to change the rules of the regime but rather to operate outside of the regime.

Similarly, certain forms of collaboration (e.g. machinery rings, see Schiller *et al.*, this volume) do not question the emphasis of the regime on the modernization of agriculture, or its productivist values. Rather, they offer an alternative to the practice of each farmer purchasing his/her own machinery. Whilst this clearly is a new form of collaboration between farmers, questioning the individualistic notion of farmers competing against each other, it leaves unquestioned many other aspects of modernized agriculture. Furthermore, some niches may occur as a means to secure the survival of farming, when the dominant regime does not offer a solution. This is the case in marginal farming areas, such as in south-east Portugal, where the extensive silvo-pastoral livestock farming system in Montado areas is under threat by both intensification and extensification trends. As a result, a niche has emerged promoting multifunctional management and the re-valorization of the quality of products, such as aromatic herbs and mushrooms, to support the survival of the Montado (see Schiller *et al.*, this volume).

However, whilst niches themselves may not (currently) strive to initiate a transition, it does not necessarily mean that they do not have a role to play in a (potential or wider) transition. Indeed, the mere existence of alternatives can encourage other networks to propose further alternatives, and each niche represents a potential for synergistic action at a later point in time.

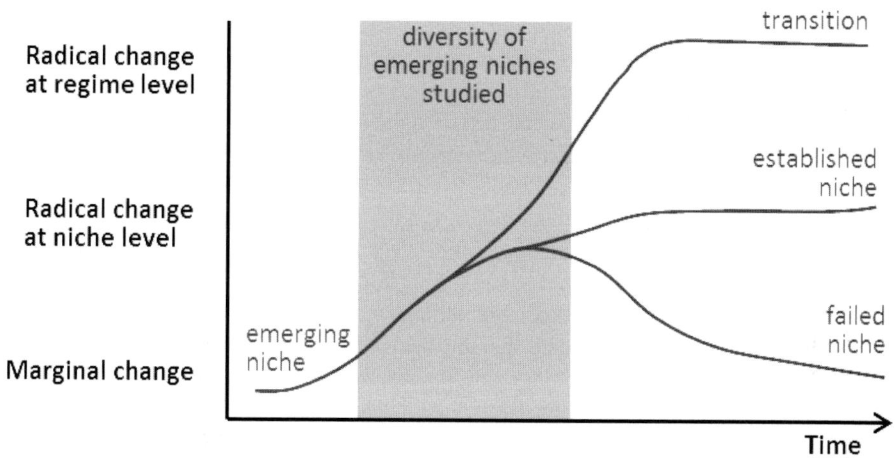

Fig. 13.1. The transformative ambition of the niche is important, as it will influence the pace and effectiveness of a potential transition at regime level. There are also niches with no ambition to change the dominant regime; wishing instead to establish themselves as a local option. Yet, even this type of niche might contribute to initiating a transition if the idea is 'picked up' elsewhere by other actors, leading to niche proliferation. (Note that lines were used in the figure for pragmatic reasons and should not imply that the trajectory is smooth and linear: niche development is usually a contested and fragmented process). (Source: authors)

Not only were the ambitions of some niches different to what was expected, a number of case studies were not initiated by entrepreneurs but co-created through targeted intervention by governments. For example, in the cases linked to high nature value farming (see Peneva *et al.*, this volume), two of the three niches studied were introduced through targeted action by respective national governments, as part of combined environmental protection and rural development initiatives. As explained above, the niche was supported by the regime itself, as a means to find solutions for farm systems in decay due to their low economic competitiveness, but still valuable from social and environmental perspectives.

This start-up process might be linked to a specificity of farming: strong policy involvement. Given the role of farmers as managers of natural resources, multifunctional agricultural policies have tended to focus on creating opportunities to enhance environmental service provision. Regarding their future development, it is unclear to what extent the fact that the niches were co-created between regime actors and niche entrepreneurs affects the longevity of the niche (e.g. once support payments are discontinued), or its potential to act in a transformative way.

Finally, some of the initiatives were distinct from niches in the sense of coordinated actors, in that they emerged from the independent actions of atomistic decision makers. For example, lifestyle farmers (see Pinto-Correia *et al.*a, this volume) follow the same trend but without coordination. Whilst they are not formally organized, they may still be considered a niche since they represent a distinct set of practices and values, and the initiative could lead to a transition at the regional level. Indeed, farmland is no longer valued solely for its production potential but also as a place of residence, for the quality of life that it offers and the attractiveness of the landscape.

These examples show that in farming, niches might differ from the ideal type within the MLP in that they may not have the ambition to transform the regime, that they may be co-created by niche-entrepreneurs and regime actors, or that they may emerge from atomistic decision makers without any formal coordination. How these differences in ambition, genesis or level of internal coordination may affect the transformative potential of a niche is unclear, and could depend on serendipity and interactions with other niches that are active in the same region.

Linking with which regime?

In the literature on transitions, the socio-technical regime refers to a "semi-coherent set of rules that guide, orient and coordinate the activities of the social groups that reproduce the various elements of socio-technical systems" and thus accounts for the stability of the system (Geels, 2011:27). Whilst the operationalization and specification of regimes has been understood as challenging (Geels, 2011), this is linked to the problem of defining the topic of analysis. The 'regime' is thus an interpretive analytical concept that invites the analyst to investigate what lies underneath the activities of the actors, such as shared beliefs, policy paradigms, visions, social expectations and norms, lifestyles of users, institutional arrangements and regulations (Geels, 2011). For each case study, it is, thus, first necessary to define the scope of the empirical topic so as to define what the 'regime' is, in the context of that particular topic. This choice should be made transparent, given that the way in which the regime is defined will determine which processes and change dynamics can be analysed, whilst at the same time obscuring others.

Whilst transformation dynamics are often driven by the tensions and contradictions within a regime, it has been pointed out that transitions often involve novel links between previously separate societal domains (Holtz *et al.*, 2008). When analysing a specific situation, it seems useful to explicitly identify which societal domains are involved, and how they are involved (Holtz *et al.*, 2008). If these domains – addressing different societal functions – are coordinated by different sets of rules, they may be labelled as separate regimes. Distinguishing between regimes might be particularly appropriate when studying potential transitions in farming (see Fig. 2.2 in Darnhofer, this volume) given the diverse social expectations that it faces, and the multifunctional nature of farming. Furthermore, in farming there is always the issue of land, which is a stable factor but which by its territorial expression, is dependent on a multiple set of regimes such as agriculture, natural resources management and real-estate. If the topic of analysis reveals the involvement of several regimes, including them will allow identification of tensions both within each regime and between regimes. It also highlights that a niche may build ties with one regime and use these ties to reinforce the pressure on another regime (Diaz *et al.*, 2013).

Including several regimes was useful in a number of case studies, thus highlighting the role of multi-regime interaction. The clear distinction between regimes was often difficult, not least due to actors linked to multiple regimes and the difficulty in defining the boundaries of a specific regime. One of the central issues is access to land and control over its management: land is a basic requirement for agricultural activities but at the same time, it is also used as property, or for recreation and nature conservation. The related regimes are, thus, necessarily intertwined in any process related with land. For example, the mainstreaming of renewable energy production on farms (Sutherland *et al.*b, this volume) was primarily the result of subsidies introduced through the energy regime, which aimed to promote energy production from renewable resources. Given that much rural land is owned by farmers, these areas provided sites for wind turbines, or to establish biogas plants and provide the necessary feedstock. This resulted in a novel interaction between the energy and agricultural regimes. Similarly, the cases related to countryside consumption (Pinto-Correia *et al.*a, this volume) show that shifts in the housing regime, and broader social trends regarding quality of life, can push an increasing number of home owners to purchase (small) farms and become 'lifestyle' or 'hobby farmers'; thus, competing for land with 'active' or 'commercial' farmers. Another important interaction is with the environmental protection regime: the Nitrate Directive, the need to increase biodiversity or protect sensitive ecosystems, creates pressures on the agricultural regime (Diaz *et al.*, 2013). The tension between the two regimes creates opportunities for many niches within which alternative – environmentally friendly – practices are promoted, both at farm level and along the food chain.

The central role of multi-regime interactions in the niches studied within the FarmPath project are linked to the multifunctionality of agriculture, and so may be expected to become a distinguishing feature of transitions in farming. This is especially the case if a territorial approach is selected (that is one that looks at the different activities and roles of a farm in its natural and social context), rather than taking a sectoral approach where the farm is seen primarily as an element along the food chain. As a result, a farmer may be engaged in food production, energy production, amenity production and environmental protection. Farmers could, thus, be considered as actors engaged in multiple regimes, making the farm a locus of tension between different demands linked to different regimes. Simultaneously, these links with different regimes offer various options on how, and how much, individual farmers engage with each regime, again reinforcing the diversity of farming and of the niches linked to farming.

Niche-regime interactions

The fact that within farming several regimes are likely to be involved in an emerging transition highlights new mechanisms for niches to influence a regime. Indeed, a niche may be able to build links with one regime, not least because their values and aims are compatible. For example, a niche promoting environmentally friendly farming practices might find a 'natural ally' in actors from the environmental protection regime (see Diaz *et al.*, 2013). However, while the niche is highly compatible with one regime, it may well be radical in relation to another regime, so that building links with that regime may be much more challenging.

Before analysing the question 'how does the niche link with the regime?' it is important to assess 'with what regime does the niche link?' Indeed, in some of the case studies presented in this book it was less a matter of niche versus regime, and more about a niche that has linked with one regime and together worked towards transforming a third regime. Whilst a niche is frequently "informed, initiated and designed in response to sustainability problems perceived in the regime" (Smith, 2007:436), this sustainability problem is also perceived by other regimes. These other regimes may make natural allies who offer resources to the niche. As Smith (2007) points out, the challenge for a niche to link with a regime is less arduous for niches which are aligned to it. In understanding the take-off phase of a transition, it might therefore be useful to pay attention to dynamics between regimes; how actors from one regime might support a niche in its efforts to transform another regime.

In terms of these dynamics, a number of processes and actor types were explored during the research process. Elzen *et al.* (2012) identify three types of 'anchoring' to describe the linking processes between a novelty and existing structures and institutions: technical, network and institutional. Technological anchoring refers to the further specification of the technology around which the niche is focused. In studying cases which were largely based on new practices and ways of working, rather than technologies per se, technological anchoring involved the development of products for market, new means of credit access, or other forms of collaboration. Technological anchoring therefore overlapped with 'network anchoring': the recruitment of new actors into the niche. Darrot *et al.* (this volume) described these partnerships as 'niche-tandems': relationships developed across multiple sub-regimes (e.g. producers working with consumers; processors enrolled in the tourism industry). These strong bonds between individuals operating in different regimes were important for creating new spaces for novelties to become niches, and for niches to develop (e.g. through securing access to markets or state support). Although anchoring typically occurred, initially, between two interest groups, Vlahos and Schiller (this volume) found that the order in which this anchoring occurred was not important; anchoring processes were in any case iterative as the niche developed. This growing interaction between members of multiple regimes led to what Lošťák *et al.* (this volume) termed 'regime grafitting', whereby the rules or expectations of one regime were transposed onto the niche's governance structure (such that during processes of institutional anchoring, regulations, rules and norms could be transferred from one regime onto another).

It is important to note that whilst a niche might not aim to transform the regime, it may engage with the regime in such a way as to create conditions that are conducive to its further development. For example, an initiative might be successful in defining an agri-environmental measure suited to its needs so as to be able to receive direct payments; a form of institutional anchoring (see Elzen *et al.*, 2012; Diaz *et al.*, 2013). With the niche successfully influencing the selection of environment, it simultaneously favourably affects the 'opportunity space' for other networks, for other niches and, thus, in effect changes the regime. Smith and Raven (2012:1030) have labelled this transformation of the selection environment 'stretch and transform empowerment', where some features of the niche are institutionalized as new norms and routines in a transformed regime. The niche is then empowered by enabling it to change its selection environment, rather than being subordinated to it. In other words, whilst the niche might not have radically changed the whole regime, it has successfully re-structured a specific aspect in ways that are favourable to itself and probably other niches that follow similar aims. This process might be particularly important in emerging transitions, as it empowers the niche and increases its competitiveness, essential characteristics for the 'take off' phase of a transition.

However, such an empowerment may face limits. Whilst the regime can create room for some niches to anchor – and as such, seem to be opening up to a transition – it might actually strive to contain the niche and the possible transition within some pre-established boundaries, so as to ensure that it does not influence the regime further (as shown in Peneva *et al.*, this volume). The Rural Development Programme under Pillar II of the CAP (a sub-regime of policy within the agricultural regime), to some extent shows such a trend: there are some dimensions of change and there is some support for rural development initiatives which comply with European public demand for a healthy countryside and environmentally friendly agriculture. However, the bulk of the agricultural policy continues to support productivist farming with no significant change in the regime.

The influence of changes in the socio-technical landscape

The term 'socio-technical landscape' refers to the wider context which influences niche and regime dynamics; it includes the technical and material backdrop that sustains society, as well as demographical trends, political ideology and societal values (Geels, 2011). Actors at niche and regime levels cannot influence the socio-technical landscape in the short term.

Given that farming is at the crossroads of a number of societal functions, a wide range of pressures from the socio-technical landscape will be encountered. For example, both climate change and the need to decarbonize energy supplies play important roles in the support for renewable energy (e.g. wind turbines and biogas plants, see Sutherland *et al.*b, this volume). Recognition of the environmental damage caused by intensive agricultural practices has led to pressure on the agricultural regime to limit certain actions, such as the amount of nitrogen fertilizer used, in order to reduce nitrate levels in surface water, thus opening a window of opportunity for a whole range of extensive farming practices (see Vlahos and Schiller, this volume). Food scares linked to industrial food production have led some consumers to seek out, and establish, alternative food networks, some of which have been supported by municipalities (see Darrot *et al.*, this volume). In all of the case studies reported in this book, landscape pressures caused by persistent problems in the regime can clearly be identified, opening windows of opportunity for niches.

The role of researchers

When studying emerging transitions, researchers often place themselves in the role of a neutral observer analysing social dynamics. It is rare that the role of research within a transition is emphasized beyond the fact that it might have contributed to the development of a new technology.

However, there have increasingly been calls for 'engaged research' (Woods, 2010; Campbell and Rosin, 2011), where researchers take a reflexive role and acknowledge how their research practices influence unfolding transitions. Others (Funtowitcz and Ravetz, 1993; Nowotny *et al.*, 2003) have pointed out that unless there is a shift in how research is done, it may no longer be able to address societal issues. Within this context, there are increasing calls for transdisciplinary research (see, for example, Pohl and Hirsch Hadorn, 2007).

In the FarmPath project, there was a clear effort to integrate transdisciplinary elements (see Pinto-Correia *et al.*b; Karanikolas *et al.*, both this volume). These efforts have highlighted not only the potential additional insights in transition processes but also the constraints imposed by the lack of experience of many researchers in such processes. They have also highlighted the challenge that implementation poses in an international project, where researchers differ in their ontological assumptions, epistemological commitments and methodological choices.

By promoting reflexivity, such research projects also promote learning by researchers and, thus, potential transformation within the agricultural research sub-regime. Indeed, research is a constituent of the agricultural regime, and thus deeply involved in co-constructing its underlying norms, logic and values, whilst at the same time (often unconsciously) shutting out alternatives (see Vanloqueren and Baret, 2009). How research is performed, thus, plays an important role in enabling the continued dominance of the regime, or enabling the emergence of a transition. This is particularly the case regarding the extent to which practitioners are included in the research process, how they are integrated, and whether methods are selected that capture empirical diversity and highlight 'outliers', rather than focusing solely on dominant practices (see Gibson-Graham, 2008). Especially in emerging transitions, transdisciplinary and participatory processes may empower local agents of change: giving them legitimacy, offering a platform where their approach can be discussed in a wider setting, and making diversity visible. Recent trends within research highlight the potential of co-design, where knowledge is developed in a complex, interactive design process with a range of stakeholders through a process of social learning (Grin *et al.*, 2010). This can lead to re-framing or a change in perspective amongst stakeholders who seek a shared problem perception, and directions for sustainable solutions (Kemp and Loorbach, 2006).

The experiences within FarmPath confirm that transdisciplinary work implies careful understanding of the dynamics within participatory processes, and active engagement by researchers. This calls for a fundamental change in researchers' attitudes, which can be challenging. Indeed, in FarmPath some researchers have felt uncomfortable with this process, partly due to a lack of training and also to their late integration into an ongoing case study (see Pinto-Correia *et al.*b, this volume).

Challenges in identifying a 'transition' in farming

The ideal or typical representation of a transition is a process that results from niche-innovations that build up internal momentum and changes at the level of the socio-technical landscape that create pressure on the regime thereby creating windows of opportunity for niche-innovations (Geels, 2011). However, Geels (2011:29) has pointed out that there are no simple causalities in transitions; instead there are "processes in multiple dimensions and at different levels which link up with, and reinforce each other ('circular causality')".

In FarmPath, the aim was not to study completed transitions but current, ongoing processes that – *a priori* – seemed to have the potential to initiate transition at a later time (emerging transitions). The challenge was to select initiatives that had the potential to lead to radical change; achieved with varying degrees of success. Indeed, when selecting potential cases, a number of issues arose around the identification of whether an initiative was in the 'take-off phase', and what would constitute a 'transition'.

Generally, it has proved challenging to identify niches involved in an emerging transition. The choice of which initiatives to select for study must be made at an early stage of the research process, and is usually based on limited information. Whilst a niche might *a priori* seem promising, upon closer analysis different aspects are revealed that indicate limitations in niche dynamics, leading to the conclusion that a take-off is unlikely. Furthermore, whether a niche will successfully navigate the take-off stage and, thus, initiate a transition, and whether that transition will be realized (or falter) can only be ascertained with hindsight.

The other challenge is linked to selecting the criteria that differentiate a transition from marginal change. Indeed, a transition is characterized by radical change at regime level. When discussing a transition, it is essential to differentiate between radical changes (fundamental shifts in system logic) and incremental changes (e.g. when the regime adapts in response to landscape pressure, or when it co-opts a niche). The challenge is that incremental versus radical change is not a binary (either/or) judgement, as one might lead to the other. In the context of emerging transitions it is not clear whether observed incremental changes may coalesce into a radical change. There is a distinct possibility that incremental changes may accumulate, but not add up to a radical change. The challenge is, then, to distinguish between radical change and those marginal changes which are part of the on-going adaptations of the regime (that is changes that do not question fundamental values, paradigms, social expectations and norms, lifestyles of users, or institutional arrangements and regulations).

The challenge of identifying transitions is partly linked to the definition of the regime, (the level of analysis). Indeed, "what looks like a regime shift at one level may be viewed merely as an incremental change in inputs for a wider regime at another level" (Geels, 2011:31). Thus, defining the regime as the 'agricultural regime' within a region may lead to the conclusion that the niche being studied cannot be reasonably expected to lead to a transition in the agricultural regime, whilst missing the fact that the niche does have a radical impact at a lower spatial scale (e.g. the horticultural sector of the region).

Another analytical challenge is linked to the fact that regimes are understood as 'dynamically stable' in that a regime undergoes change constantly whilst retaining its fundamental characteristics. Thus, the fact that the regime underwent (radical) change may, or may not, be linked to the activities of the niche. To ascertain the impact of a niche it would be helpful to assess the counterfactual situation: how would the regime have evolved if the niche had not been active, or had been less successful? However, it is not only the regime that changes over time, the niche does too: through the influence of interactions with the regime, and through interactions with other niches, such as when enlarging its network. Assessing the impact of a niche upon a regime, thus, raises the analytical challenge of disentangling the complex dynamics of an evolving niche, an evolving regime and their co-evolving interactions.

A challenge which could be particularly pronounced when analysing transitions in farming is how to take into account the diversity of practices within a region. Indeed, identifying a radical change in values and practices is easier if the 'dominant' practices are homogeneous (e.g. transportation by automobiles), or if there is one dominant practice with a number of minor practices (e.g. in energy production), particularly if these practices rely on large investments in infrastructure (e.g. roads, power plants). However, in farming the 'mainstream' may be more an artefact of the rhetorical power of regime actors (e.g. the Chamber of Agriculture) than a faithful description of farmers' practices. Research based on an actor-orientation, such as on farming styles (van der Ploeg, 2000), or on rural development (van der Ploeg *et al.*, 2000), has shown that farmers are not guided in a deterministic way by structural constraints such as subsidies, markets or their natural environment but actively mediate their impact, resulting in a variety and diversity of on-farm practices. Thus, given farmer agency and the dependence of farming on the natural and cultural environment, a uniform 'transition' is unlikely. Yet, it is debatable whether a shift in the proportion of farms using a particular practice (e.g. direct marketing, energy production), or a shift towards a higher level of diversity, is sufficient to be labelled a transition.

The multifunctionality of farming (the fact that farms are at the crossroads of several societal functions, and thus may be involved in several regimes) also adds to the challenge. Indeed, at farm-level and within a region, several transitions may be emerging. For example, in Aberdeenshire (UK), a renewable energy transition is occurring at the same time as countryside consumption is becoming more common (see Pinto-Correia *et al.*a, and Sutherland *et al.*, both this volume). Given that farming is influenced by these multiple developments, it can prove challenging to clearly identify one transition, in one sector. This is especially the case if several 'marginal' changes are occurring in each sector, each of which – when considered in isolation – does not merit the term 'radical' and, thus, is not a transition in that regime. Taken together, however, these changes do significantly alter the dynamics at regional level. In addition, transition theory makes it possible to understand that the changes may be not within the regimes themselves but in a shift in the dominant role of one regime over another. Referring to the example of farms in one area mentioned above, the energy regime or the real estate regime may become more relevant than the agricultural regime. From an analytical perspective, the challenge is to define regime(s) in such a way as to adequately capture these dynamics.

Conclusion

Transitions are about "relatively rare, long-term macro-changes" (Geels, 2011:38). Since they do not occur frequently, it is difficult to identify suitable case studies, and even more so if the focus is on emerging transitions whose future trajectory is unknown. However, studying and engaging with emerging transitions is particularly relevant, as it may lead to the empowerment of niches which have the potential to transform dominant practices, and therefore contribute to increasing the sustainability of societal practices.

The purpose of applying the MLP to farming was to identify mechanisms and processes that strengthen or weaken the transformative potential of niches, providing insights that other analytical perspectives do not reveal. This perspective can bring new knowledge that can support the design and targeting of public policies when they seek to engender transformative change. However, the work of the FarmPath project has highlighted several challenges when applying the MLP in an agricultural setting. This is primarily due to the defining characteristics of the agricultural sector: the diversity of practices, even within a region; the spatial nature of opportunities; the multifunctional role of farms; the public nature of the 'goods' produced; and the high level of policy involvement. In this context, the emergence of niches and the dynamics of niche-regime interactions are likely to be different compared to other sectors, making the study of emerging transitions even more challenging. To address this challenge, the analytical focus on the anchoring process has been productive, highlighting that niche-regime(s) interactions are relevant, as well as the interactions between regimes.

The context-dependency of many processes of change in agriculture has also made clear that it is highly unlikely that an innovation (e.g. a technology, a production method or the organization of a value chain) can be 'up-scaled', in that it cannot serve as a blueprint in another context, even if it may serve as inspiration for similar processes elsewhere. Nor can we expect an innovation to spread uniformly within a region. A transition to sustainability in agriculture is thus unlikely to be a homogeneous process, with clearly defined characteristics that can be captured through statistical trends. It is more likely to be a diffuse, fuzzy intermingling of diverse, co-evolving niches and trends, out of which radical change emerges.

The diversity within farming, and the fact that farms are at the crossroads of several regimes, also make it likely that within a region there will be a multiplicity of niches, initiatives and projects. None of these may be large enough to muster the transformative energy needed to single-handedly engender a transition. However, this does not justify discarding these 'small' niches, as they may network with others, creating synergies that, together, fundamentally alter the dynamics within a region.

References

Campbell, H. and Rosin, C. (2011) After the 'organic industrial complex': An ontological expedition through commercial organic agriculture in New Zealand. *Journal of Rural Studies* 27, 350–361.

Darnhofer, I. (2015) Socio-technical transitions in farming. Key concepts. In: Sutherland, L-A., Darnhofer, I., Wilson, G.A. and Zagata, L. (eds) *Transition Pathways towards Sustainability in Agriculture: Case Studies from Europe.* CABI, Wallingford, UK, pp. 17–32.

Darrot, C., Diaz, M., Tsakalou, E. and Zagata L. (2015) 'The missing actor': Alternative agri-food networks and the resistance of key regime actors. In: Sutherland, L-A., Darnhofer, I., Wilson, G.A. and Zagata, L. (eds) *Transition Pathways towards Sustainability in Agriculture: Case Studies from Europe.* CABI, Wallingford, UK, pp. 143–156.

Diaz, M., Darnhofer, I., Darrot, C. and Beuret, J-E. (2013) Green tides in Brittany: What can we learn about niche-regime interactions? *Environmental Innovation and Societal Transitions* 8, 62–75.

Elzen, B., Leeuwis, C. and van Mierlo, B. (2012) Anchoring of innovations: Assessing Dutch efforts to use the greenhouse effect as an energy source. *Environmental Innovation and Societal Transitions* 5, 1–18.

European Union (2012) *The Common Agricultural Policy. A Story to be Continued.* OPOCE (Office for Official Publications of the European Communities).

Funtowitcz, S. and Ravetz, J. (1993) Science for the post-normal age. *Futures* 25, 739–755.

Geels, F.W. (2011) The multi-level perspective on sustainability transitions: Responses to seven criticisms. *Environmental Innovation and Societal Transitions* 1, 24–40.

Gibson-Graham, J.K. (2008) Diverse economies: Performative practices for 'other worlds'. *Progress in Human Geography* 32, 613–632.

Grin, J. (2010) Modernisation processes in Dutch agriculture, 1886 to the present. In: Grin, J., Rotmans, J. and Schot, J. (eds) *Transitions to Sustainable Development. New Directions in the Study of Long Term Transformative Change*. Routledge, New York, pp. 249–264.

Grin, J., Rotmans, J. and Schot, J. (eds) (2010) *Transitions to Sustainable Development. New Directions in the Study of Long Term Transformative Change*. Routledge, New York.

Holtz, G., Brugnach, M. and Pahl-Wostl, C. (2008) Specifying 'regime' – A framework for defining and describing regimes in transition research. *Technological Forecasting & Social Change* 75, 623–643.

Karanikolas, P., Vlahos, G. and Sutherland, L-A. (2015) Utilising the multi-level perspective in empirical field research: Methodological considerations. In: Sutherland, L-A., Darnhofer, I., Wilson, G.A. and Zagata, L. (eds) *Transition Pathways towards Sustainability in Agriculture: Case Studies from Europe*. CABI, Wallingford, UK, pp. 51–66.

Kemp, R. and Loorbach, D. (2006) Transition Management: A reflexive governance approach. In: Voß, J.P., Bauknecht, D. and Kemp, R. (eds) *Reflexive Governance for Sustainable Development*. Edward Elgar Publishing, Cheltenham, UK.

Lošťák, M., Karanikolas, P., Draganova, M. and Zagata, L. (2015) Local quality and certification schemes as new forms of governance in sustainability transitions. In: Sutherland, L-A., Darnhofer, I., Wilson, G.A. and Zagata, L. (eds) *Transition Pathways towards Sustainability in Agriculture: Case Studies from Europe*. CABI, Wallingford, UK, pp. 157–170.

Marsden, T. (2013) From post-productionism to reflexive governance: Contested transitions in securing more sustainable food futures. *Journal of Rural Studies* 29, 123–134.

Marsden, T. and Sonnino, R. (2008) Rural development and the regional state: Denying multifunctional agriculture in the UK. *Journal of Rural Studies* 24, 422–431.

Nowotny, H., Scott, P. and Gibbons, M. (2003) 'Mode 2' revisited: The new production of knowledge. *Minerva* 4, 179–194.

Peneva, M., Draganova, M., Gonzalez, C., Diaz, M. and Mishev, P. (2015) High nature value farming: Environmental practices for rural sustainability. In: Sutherland, L-A., Darnhofer, I., Wilson, G.A. and Zagata, L. (eds) *Transition Pathways towards Sustainability in Agriculture: Case Studies from Europe*. CABI, Wallingford, UK, pp. 97–112.

Pinto-Correia, T., Gonzalez, C., Sutherland, L-A. and Peneva, M. (2015a) Lifestyle farming: Countryside consumption and transition towards new farming models. In: Sutherland, L-A., Darnhofer, I., Wilson, G.A. and Zagata, L. (eds) *Transition Pathways towards Sustainability in Agriculture: Case Studies from Europe*. CABI, Wallingford, UK, pp. 67–82.

Pinto-Correia, T., McKee, A. and Guimarães, H. (2015b) Transdisciplinarity in deriving sustainability pathways for agriculture. In: Sutherland, L-A., Darnhofer, I., Wilson, G.A. and Zagata, L. (eds) *Transition Pathways towards Sustainability in Agriculture: Case Studies from Europe*. CABI, Wallingford, UK, pp. 171–188.

Pohl, C. and Hirsch Hadorn, G. (2007) *Principles for Designing Transdisciplinary Research*. Oekom, Munich, Germany.

Renting, H., Oostindie, H., Laurent, C., Brunori, G., Barjolle, D., Jervell, A., Granberg, L. and Heinonen, M. (2008) Multifunctionality of agricultural activities, changing rural identities and new institutional arrangements. *International Journal of Agricultural Resources, Governance and Ecology* 7, 361–385.

Schiller, S.R., Gonzalez, C. and Flanigan, S. (2015) Collaboration in agriculture. In: Sutherland, L-A., Darnhofer, I., Wilson, G.A. and Zagata, L. (eds) *Transition Pathways towards Sustainability in Agriculture: Case Studies from Europe*. CABI, Wallingford, UK, pp. 83–96.

Slee, B. and Pinto-Correia, T. (2015) Understanding the diversity of European rural areas. In: Sutherland, L-A., Darnhofer, I., Wilson, G.A. and Zagata, L. (eds) *Transition Pathways towards Sustainability in Agriculture: Case Studies from Europe*. CABI, Wallingford, UK, pp.33–50.

Smith, A. (2007) Translating sustainabilities between green niches and socio-technical regimes. *Technology Analysis and Strategic Management* 19, 427–450.

Smith, A. and Raven, R. (2012) What is protective space? Reconsidering niches in transitions to sustainability. *Research Policy* 41, 1025–1036.

Smith, A., Stirling, A. and Berkhout, F. (2005) The governance of sustainable socio-technical transitions. *Research Policy* 34, 1491–1510.

Sutherland, L-A., Peter, S. and Zagata, L. (2015) On-farm renewable energy. In: Sutherland, L-A., Darnhofer, I., Wilson, G.A. and Zagata, L. (eds) *Transition Pathways towards Sustainability in Agriculture: Case Studies from Europe*. CABI, Wallingford, UK, pp.127–142.

van der Ploeg, J.D. (2000) Revitalizing agriculture: Farming economically as starting ground for rural development. *Sociologia Ruralis* 40, 497–511.

van der Ploeg, J.D. and Roep, D. (2003) Multifunctionality and rural development: The actual situation in Europe. In: van Huylenbroeck, G. and Durand, G. (eds) *Multifunctional Agriculture*. Ashgate, Aldershot, UK, pp. 37–53.

van der Ploeg, J.D., Renting, H., Brunori, G., Knickel, K., Mannion, J., Marsden, T., de Roest, K., Sevilla-Guzmán, E. and Ventura, F. (2000) Rural development: From practices and policies towards theory. *Sociologia Ruralis* 40, 391–408.

Vanloqueren, G. and Baret, P. (2009) How agricultural research systems shape a technological regime that develops genetic engineering but locks out agroecological innovations. *Research Policy* 38, 971–983.

Vlahos, G. and Schiller, S. (2015) Transition processes and natural resource management. In: Sutherland, L-A., Darnhofer, I., Wilson, G.A. and Zagata, L. (eds) *Transition Pathways towards Sustainability in Agriculture: Case Studies from Europe*. CABI, Wallingford, UK, pp.113–126.

Woods, M. (2010) Performing rurality and practicing rural geography. *Progress in Human Geography* 34, 835–846.

Chapter 14

Conclusions

L-A. Sutherland[1], L. Zagata[2] and G.A. Wilson[3]

[1]*James Hutton Institute, Aberdeen (lee-ann.sutherland@hutton.ac.uk);* [2]*Czech University of Life Sciences, Prague;* [3]*Plymouth University*

Introduction

The primary purpose of this book has been to assess the mechanisms involved in enabling sustainability transitions in European agriculture. The research focused on *emerging transitions*: 'transitions in the making', with the potential to have major impacts on mainstream farming practices. This emphasis was intentionally normative, specifically oriented towards assessing transition processes which aim to increase the *sustainability* of agriculture in Europe, by optimizing the resources and opportunity sets evident at regional level. The need for increased sustainability has been described in the introduction to this book (Sutherland *et al.*a, this volume). Challenges include the growing and varied demands on agricultural systems by society, including food security, safety and quality; climate change and competition for natural resources; economic recession and the subsidy dependence of EU agricultural production systems; rural outmigration and land abandonment in remote areas; intensification and counter-urbanization in peri-urban areas; and a perceived shortage of young people working in agriculture. In this chapter we highlight the cross-cutting themes from the research, emphasizing in particular the implications for sustainability transitions and the role of young people in innovation processes. We conclude by identifying future research needs, specifically in relation to the utility of the multi-level perspective (MLP), for the study of agricultural transitions.

One of the key contributions of this research has been the operationalization of the MLP for use in empirical field research. Previously, the MLP had primarily been used to assess historic, completed transitions (see Karanikolas *et al.*, this volume) Analysis, thus, appears to have relied on document review but methods are rarely discussed[1]. In order to assess emerging transitions, empirical field research was required. The particular challenges of operationalizing the MLP, and the methods adopted, have been discussed by Karanikolas *et al.* and Pinto-Correia *et al.*b (both this volume), with theoretical implications addressed by Darnhofer *et al.* (this volume). These chapters demonstrate that

[1] Geels and Schot (2010) describe their methods broadly as a form of 'process theory', where narrative is generated through theoretically informed analysis of events. However, the formal description of data sources, analytical techniques and methodological limitations characteristic of most empirical field research studies, are noticeably absent from the MLP literature.

use of the MLP is not straightforward for emerging transitions: the boundaries of the niche, regime and landscape factors are not easily defined. Processes and dominant players, which are clearly evident in retrospect, are less so in mid-transition and it is difficult to identify *a priori* those niches which are likely to lead to future regime change. In addition, in order to derive conclusions which are useful for influencing and intervening in emerging transitions, a transdisciplinary approach was adopted whereby researchers collaborated with lay participants and stakeholders.

Adding to the challenge of operationalization is the 'special case' represented by the agricultural sector in transition studies. Marsden (2013) identifies two reasons for this: the inherently spatial nature of agricultural systems (based on land) and the high level of policy intervention in agricultural systems (related to public goods provision). In Chapter 3 (this volume), Slee and Pinto-Correia describe the regional differentiation that is inherent in European agriculture, borne out in the variation in regional opportunities demonstrated in subsequent case study chapters (Chapters 5 to 11), as well as in the types of intervention encountered (see Chapter 12[2]). The extensive role of government policy in the progression (and in some cases initiation) of sustainability initiatives is also evident throughout the case studies. However, our analysis also suggests a third related distinction: the multifunctional nature of the agricultural sector. The diverse functions expected of the agricultural sector lead to extensive interactions with other sectors, including energy, tourism, real estate and nature conservation, in addition to the range of consumption, marketing and distribution functions expected of what is now often termed the 'agri-food system'.

The role of technology in transition processes may also be distinctive in the agricultural sector. Although the level of mechanization does not distinguish agriculture from other industries, it may be unique in terms of the extent of public concern over, and government intervention in, the impacts of this mechanization on the structure of the industry. Impacts can affect the environment and other service provision such as food quality, amenity, and maintenance of cultural heritage. As a result, sustainability transitions in agriculture are less likely to depend on new technologies, instead involving 'soft' forms of transition (e.g. through new governance measures, alternative marketing channels, 'retro-innovation' and relocalization). These special characteristics offer an opportunity to further develop the MLP for use in assessing transition processes both within and beyond the agricultural sector.

Emerging themes in sustainability transitions

Technological bias within the MLP

The MLP was originally developed to assess technology-based transition processes. Based on thinking from evolutionary economics and science and technology studies (Geels and Schot, 2010), these foundations are reflected in the emphasis on 'socio-technical systems', which place the development of technologies as central to transition processes, whilst recognizing the broader systems in which they originate and evolve. Definitions of

[2] Although the interventions identified were highly differentiated, reflecting regional differences, the goals identified for agriculture were remarkably similar across the sites.

'technological niches' are consistently traced back to Rip and Kemp (1998), who emphasize that niches are 'configurations which work': the 'technology' aspect of the niche is not limited to physical equipment or machinery but can include new production methods and ways of working, such as short supply chains. However, to date socio-technical niches assessed utilizing the MLP have been almost entirely based around specific physical technologies. The exceptions have primarily been studies within the agricultural sector, particularly in relation to organic farming (e.g. Belz, 2004; Smith, 2006, 2007). Similarly, in this research we have focused primarily on the configurations which work (such as farmers' markets, local certification and collaborative actions).

In assessing the case studies presented in this book, the findings suggest a 'technological bias' within the MLP: conceptual assumptions that are the result of basing empirical research primarily on physical technology-based transition processes. In this research, only one of the chapters focused around a physical technology-based niche: on-farm renewable energy production (Sutherland *et al.*b, this volume). This cluster of cases was the most straightforward to assess in terms of applying the MLP concept. The cases demonstrated clear evidence of niche development and anchoring within the three regional contexts, with 'translation' of the original purposes of the technologies in response to the 'window of opportunity' presented by the emerging climate change agenda. However, in cases without a new physical technology as the focal point for a transition process, it can be difficult to identify the boundaries of a niche; the processes of anchoring by the niche; or to differentiate between incremental and transitional regime change. This was particularly true in the case of lifestyle farming (Pinto-Correia *et al.*a, this volume), where the increase in consumption-oriented occupancy of rural land was clearly evident but difficult to define. Although physical technologies develop and take on new forms and purposes, associated developments are much simpler to follow than new market arrangements or forms of collaborative action.

Physical technology-based innovations are also more straightforward to support. Funding research and development is common practice for governments and industries; both were important for the development of renewable technologies and the subsequent mainstreaming of production. Sutherland *et al.*b (this volume) suggest that in the British and German cases, governments viewed renewable energy production both as an economic development tool and as a means of meeting their climate change mitigation objectives. Targets for production could be clearly identified and assessed, achieving multiple goals. In contrast, supports for organizational innovations, such as network development, are much more difficult to quantify in terms of outputs and relative success. As a result, it may be easier for niches to access funding support for technology-based transition processes.

It should also be noted, however, that although one of the strengths of the MLP is the attention it draws towards technology, the role of physical technologies other than those characterizing socio-technical niches is not clear. Technologies could be considered either as part of the socio-technical landscape, or embedded within identified regimes. However, the central place of physical technologies within the MLP led the FarmPath researchers to consider the role of technologies in agricultural transition processes more carefully than might otherwise be the case (through assessing which technologies are involved). In many cases, the niche involved the revival, or revitalization, of a historic method or technology.

Lošťák *et al.* (this volume) discusses *retro-innovation*: the reformulation of traditional methods such as drying fruit and producing food, which were reinvented and modernized to appeal to consumers of local food. High nature value farming (HNVF) (Peneva *et al.*, this volume) similarly revived traditional land management practices in remote areas; alternative agri-food networks (AAFN) (Darrot *et al.*, this volume) were also, in part, a return to the shorter marketing chains of the past. These retro-innovations are different from the rediscovery of old technologies identified as 'market niches' by Markhard and Truffer (2008) in that the associated niches are primarily oriented towards new ways of working (such as networks and collaboration), rather than marketing the rediscovered technology *per se*. This combination of historic technologies and practices, with new markets and networks, is an important source of innovation in the agricultural sector.

Enabling technologies: technologies not central to the niche itself but essential to its establishment and proliferation, were also clearly important in the assessed cases. Information technology (IT), in particular, was important for connecting urban consumers with rural producers in the establishment of AAFN (Darrot *et al.*, this volume), and for enabling home working and expansion of commuter catchments in the development of lifestyle farming (Pinto-Correia *et al.*a, this volume). Use of cutting edge software facilitated the coordination of machinery and equipment sharing across wide geographic regions in the mainstreaming of machinery rings but the lack of these programmes was identified as a limiting factor to collaboration between farmers in the Regionalwert AG case (Schiller *et al.*, this volume). The research, thus, represents a first step in identifying the role of these enabling technologies within the MLP. In the agricultural sector, where production is lodged in physical space, IT access may be particularly important for remote rural regions where access has historically lagged behind that in urban and peri-urban areas.

The role of markets

The role of markets in transition processes has received considerably less attention than technology within the MLP construct, where markets are generally considered part of the incumbent regime. Elzen *et al.* (2012), in their conceptualization of anchoring, recognize the importance of markets but include them as a subset of institutional anchoring ('economic institutional anchoring'), rather than granting them a separate category in their own right. Based on the analysis in this book, we suggest that markets are more embedded in the MLP conceptualization than is currently recognized.

In the cases studied here, the role of markets was clearly important to the long-term success of many of the initiatives. For example, the establishment of viable markets was essential to the recruitment of mainstream agri-food chain actors into AAFN (Darrot *et al.*, this volume). Local certification schemes, similarly, depended on the economic viability of selling through these labels, in order to retain farming participants (Lošťák *et al.*, this volume). Market supports for renewable energy made it economically viable for both farm and corporate investors (Sutherland *et al.*b, this volume). In addition, open land markets and the buying power of incomers have been essential for the spread of lifestyle land management (Pinto-Correia *et al.*a, this volume). However, lack of markets was a particular problem associated with initiatives supporting public goods, such as HNVF (Peneva *et al.*, this volume) and natural resource management (Vlahos and Schiller, this volume). As Vlahos and Schiller point out, the socio-technical transitions literature tends to focus on market-based solutions to sustainability issues. Although it is recognized that niches initially require protection from market forces (Geels, 2007), little consideration is given to the development of niches for which there may, or may not, be a functioning market (for example, transitions associated with public rather than private goods). These cases are addressed further in the following section.

The implicit expectation of market penetration by niches, as part of the process of regime change, may reflect the centrality of physical technologies within the MLP: technology-centric transitions almost inevitably include market penetration and integration because these physical technologies are clearly marketable products. Within the MLP, markets represent central organizing forces for regime actors, reinforcing their power and position, and establishing the value of new and existing technologies and rules (such as creating 'negative' lock-ins; see Wilson, 2012). Transition is, thus, conceptualized as occurring when the incumbent regime becomes unable to respond to a landscape pressure, leading to an opportunity for existing niches to fill this gap. In the typical studies of transition processes, this leads to new market opportunities for the technologies developed and supported by niche actors (such as technology substitution pathways). Uncritical use of the MLP could lead to reliance on commercial market forces and, in its application to sustainability problems, a bias towards identification of 'technology fix' solutions (rather than new organizational forms or collaborative actions). The problems caused by market forces, such as economies of scale leading to environmental degradation and local job losses, remain unchallenged.

The cases presented in this book also raise the question: 'which markets are relevant?' Whilst it is recognized that new innovations often emerge outside, or on the periphery of a regime (Geels and Schot, 2010), studies utilizing the MLP typically focus on niche integration/transition into a single dominant regime. In the case studies in this book, some of the markets are entirely outside of the agricultural sector (for example in the case of renewable energy production, farm-produced renewable energy is sold back into the energy sector through the electricity grid, Sutherland *et al.*b, this volume). Even in the cases where the aim included food production (e.g. AAFN and local certification schemes), it is notable that these initiatives do not typically set out to transform, or anchor into, the markets of the incumbent agricultural regime. Instead, they focus on developing new, 'autonomous' (arguably hybrid) markets, anchored in multiple sectors (such as food and tourism). Mainstream agricultural marketing channels, meanwhile, appear to have remained intact. Indeed, as Darrot *et al.* (this volume) argue, autonomy was the goal of the AAFNs studied, rather than regime change. Conventional agricultural markets, thus, appear to be problems which need to be addressed through transitions towards sustainability, rather than being part of the solution.

The cases, thus, also demonstrate increasing market competition in rural areas over resource access. In the lifestyle farming and on-farm renewable energy production chapters (Pinto-Correia *et al.*a and Sutherland *et al.*b, both this volume), commercial farmers face growing competition for agricultural land. This plays out differentially, reflecting regional characteristics and the type of production involved. In the renewable energy cases focused on biogas production (Germany and the Czech Republic), where slurry and maize were also important to production, farmers retained their competitive advantage. However, in the UK case of wind energy production, where land is the sole 'agricultural' resource required, corporations are proving more able than farmers to withstand the growing economic risks of turbine development, and are increasingly dominating the transition process (see Sutherland *et al.*b, this volume). In the lifestyle farming chapter (Pinto-Correia *et al.*a, this volume), competition for land was noted as problematic in the Scottish case where lifestyle farming is leading to increased property values, which in turn restrict opportunities for young farmers. However, in the Portuguese case, lifestyle farming occurs on otherwise under-utilized land, thereby combatting land abandonment (see Pinto-Correia *et al.*a, this volume). Although it has long been recognized that farmers are no longer the dominant actors in many rural areas of Europe (Wilson, 2007), the analysis in this book brings to light concerns about the impact of open land markets on who manages natural resources like land, and for what purposes. It also raises the question of winners and losers in transition processes, an issue that Geels and Schot (2010) recognize is not typically assessed by MLP studies.

The role of policy

The public goods provided through the agricultural sector represent the justification for substantial policy investment. With the notable exception of Pinto-Correia *et al.*a (this volume), most of the cases studied in this book have had a significant amount of state intervention. Within the MLP, policy is constructed as part of the incumbent regime. Geels (2004) proposes that laws, regulations and standards represent 'regulative' rules: one of three types of rules embedded in regimes (cognitive, regulative and normative). The achievement of policy support for niches is also identified by Elzen *et al.* (2012) as an important element of institutional anchoring.

As suggested in the previous section, policy interventions are particularly important to support the protection of public goods. Renewable energy production was supported in order to address climate change issues; HNVF to increase biodiversity and population retention in remote areas; local certification schemes and AAFN, in part, to reduce the environmental impact of agriculture; and supports to natural resource management were aimed directly at improving water quality. However, as a characteristic of the frequently negative 'lock-in' of the dominant regime, existing policies can also work against sustainability transitions. Lošťák *et al.* (this volume) describe how food processing regulations, designed for large-scale processing, limited the development of locally certified products in the Bulgarian case study. They also describe the mismatch between new forms of governance at the regional level and existing governance actions at the national level. Slee and Pinto-Correia (this volume) suggest that policy supports emanating from the EU level are inherently inconsistent, with the majority of subsidies arguably targeted at productivist agriculture (CAP Pillar 1), and a minority aimed at addressing the failures of this system through Pillar 2. The policies impacting on the niche are also not necessarily set within a single sector. Sutherland *et al.*b (this volume) describe the disconnection between agriculture and energy sector supports and objectives: farm diversification grants for renewable energy production have been dwarfed by the price supports provided through the energy sector. The result is the loss of opportunities to realize long-term rural development aims through farm-based production, in favour of large-scale renewable energy developments that contribute far less to community development (see also Sutherland and Holstead, 2014). Although the need for 'joined up' cross-sectoral policies is well known, the case studies clearly demonstrate that this is far from being achieved.

Owing to the emphasis within the MLP on physical technology-based transition processes, which can be supported by traditional state research and development investments, the means of supporting the sorts of softer transition processes characteristic of sustainability initiatives in the agricultural sector are less developed. The research reported in this book represents an important step in that direction. Lošťák *et al.* (this volume) describe 'regime graffiting', where some elements of the governance structures from tourist and other regimes were transferred onto an 'agricultural' niche, ultimately leading to changes in agri-food sector governance (for example, governance mechanisms from the tourism sector were adopted in the agricultural sector, evidenced by the establishment of Czech national bodies aimed at coordinating regional labelling schemes under one umbrella). Cross-sectoral engagement at niche level can lead to institutional anchoring that alters the policy support structures of the agricultural regime.

National and regional governments also initiate niche establishment. For example, two of the three HNVF 'niches' were initiated by regional governments with funding from national government agencies, whereas the third was initiated by local farmers working with environmentalists. The two different sources pose different challenges for niche development. In conceptual terms, niches which were created 'top-down' were, at the time of their creation, already institutionally anchored in the form of state support. Those niches created by 'bottom-up' processes struggled to achieve access to funding support, often had to reformulate their purposes and legal status in order to secure these, and as a result, they jeopardized the stability of their original orientation. In contrast, niches developed from the top down retained their clear focus but struggled to develop local networks with associated visions and actions. State-led development of collaborative networks is particularly difficult in former Soviet countries, where there is cultural resistance to state mandated cooperation (Peneva *et al.*, this volume). Although different types of niches required different types of support (bottom-up networks to anchor into the regime and top-down to form coherent niches) in both cases, these networks took considerable time to develop (Lošťák *et al.*, this volume, estimate 5 to 10 years). This issue needs to be recognized, to ensure sufficient duration of government investment.

The particular challenge in facilitating sustainability transitions in the agricultural sector is that the process pits markets for food and fibre against artificially created markets for public goods. Where technology-based cases are the focus, state intervention can develop new markets (e.g. for renewable energy) through state mandated targets, with the long term aim of full market integration. For less marketable public goods, mimicking market conditions (such as using grants and income support for HNVF) can lead to successful up-take but the long-term sustainability of such subsidy dependence is questionable.

Recently, European governments have explored concepts of *payments for ecosystem services*, partly in an attempt to shift some of the burden of this responsibility. In the case study in Mangfall Valley, Germany (Vlahos and Schiller, this volume) the city of Munich opted to secure the quality of its water by subsidizing conversion to organic farming. This was effective in terms of water quality, but raised different challenges, owing to the disconnection between the purpose of the subsidy (reduction of water pollution) and the measure supported (new production techniques associated with organic farming). Poor understanding of the latter led to other problems in associated farming operations, such as reduced livestock welfare. Payment for ecosystem services is an option for states looking to reduce dependence on subsidies for the production of public goods, but further exploration of its utility and unintended consequences is required.

The role of networks

Within the MLP, the concept of networks is embedded in the conceptualization of transition processes: niche actors working together to develop novelties within a niche (Geels and Schot, 2007). Network anchoring, simply defined as changes in the network of actors who produce, use or develop the novelty, is identified as one of Elzen *et al.*'s (2012) three primary forms of anchoring, and could be considered the primary means in which the agency of actors is considered within the MLP. Network development is a primary component of the formation and development of most of the cases studied in this book.

Distinctions between bridging and bonding social capital are well recognized in the academic literature, as is the role of networks in rural development (Wilson, 2010). Bonding social capital (the connection between individuals with similar objectives) and bridging social capital (the ability to make connections to individuals with different objectives) are viewed as important in enabling change processes (Pretty, 2003). Although niches might be expected to arise initially from bonding social capital, what is evident from the cases studied here is that almost all of the niches were formed through bridging social capital. For example, all three of the local certification schemes described by Lošťák *et al.* (this volume) were formed by actors working collaboratively across the production chain (such as farmers and food processors, wine retailers and processors). All three of the collaborative actions described by Schiller *et al.* (this volume) were organized by advisors or environmental activists working in collaboration with farmers; and in all three renewable energy cases, farmers worked with engineers to develop renewable energy technologies (Sutherland *et al.*b, this volume). Darrot *et al.* (this volume) term these partnerships 'niche tandems': individuals from different sub-regimes or regimes working together towards a shared vision.

This dependence on bridging social capital has implications for niche development. We suggest that niches founded on bridging social capital will be inherently novel but also less stable than niches initially formed from bonding social capital. For 'bridging' niches, the initial vision for the niche is at least two-fold; for example, securing good prices for farmers whilst achieving the environmental aims of activists. These dual objectives can lead to conflicts, which are further complicated as the niche anchors. For example, Schiller *et al.* (this volume) describe how collective buying power is a benefit for members who purchase inputs through a machinery ring (e.g. farmers) but this can result in prices and income being driven down for other members (e.g. fuel suppliers). In some cases, the response of niche participants is to resist further bridging efforts and retreat into bonding social capital. Lošťák *et al.* (this volume) discuss two cases where once 'successful' local certification schemes, originally oriented towards regional economic development, preservation of heritage varieties and marketing of local products, later became insular and focused on meeting the needs of their own members. As such, bridging social capital had initially been strong but the connections between different groups became exclusive, restricting the development of further bridges outside of the established group.

Managing diverse members and their resources generally requires formalization of niche activities (into NGOs, for example, or legally recognized businesses). Formalization is necessary to handle the range of actors and their disparate needs (such as the growing volume of clients for machinery rings, Schiller *et al.*, this volume). Formalization is also often explicitly required to access government support (i.e. institutional anchoring): Darrot *et al.* and Lošťák *et al.* (both this volume) describe Czech cases where formalization of an NGO was necessary to access government grants. This need to formalize in response to specific state requirements further distorts the original vision of niche actors. Whilst some niches flourish under this expansion of aims, Peneva *et al.* (this volume) question what happens when the final goal of the network is achieved and, as a result, momentum may be lost. State support for these initiatives could, thus, usefully focus on developing networking and management skills (e.g. through training), and fostering interactions between different types of actors at local and regional levels (such as engaging both farmers and other rural residents in LEADER-style approaches).

An important finding of this research is that in some cases, transition processes can occur without overt network development. As Pinto-Correia *et al.*a (this volume) demonstrate, lifestyle farming developed in response to larger societal trends towards self-provisioning, appreciation of rural amenities and the 'experience economy', whilst open land markets enabled individuals to purchase peri-urban land. Anchoring occurred in the form of input supply businesses re-orienting their sales strategies towards new consumers but there was no associated social movement in two of the three cases. In MLP terms, landscape pressures directly led to an overlap between real estate and agricultural sectors, without clear network formation or anchoring.

Innovators in the agricultural sector

One of the underlying research questions addressed throughout the FarmPath project was the role of young farmers in sustainability transitions. There is widespread concern in European policy circles about the perceived shortage of young farmers in agriculture, due to their role as innovators in agricultural systems (DGIP, 2012). Evidence from the FarmPath project in support of this contention was mixed. Young people were found to be important instigators in the case studies involving AAFN and local certification schemes. In some cases, such as Regionalwert AG (Vlahos and Schiller, this volume), support for young farmers was the primary purpose of the initiative. However, young people played no appreciable role in the up-take of lifestyle farming, or on-farm renewable energy production. Our findings, thus, suggest that young farmers cannot be recognized as the main source of innovation in sustainability transition processes. Their engagement in the initiatives resulted from a combination of factors, among which the economic attraction of farming and a desire to live in the countryside were most important. Barriers become more pronounced in initiatives that require the purchase of agricultural land or substantial financial investments, and are usually connected to the economic viability of holdings, employment generated, adequacy of income, availability of housing and the quality of rural infrastructure.

In EU support programmes, young farmers are often conflated with new entrants. There was considerable evidence from the FarmPath research that new entrants were important sources of innovation. For example, Lošťák *et al.* (this volume) describe how new entrants created an innovative certification scheme for food products, using potential that had remained unrecognized by local farmers. Similarly, Pinto-Correia *et al.*a (this volume) emphasize the role of urban actors who purchased properties in the countryside in order to fulfil rural lifestyle aspirations, and used this acquired land in new, and often innovative, ways. However, there was also considerable debate about what constitutes a 'new entrant'. Although it was clear that new entrants were not necessarily 'young' (i.e. new farmers under the age of 35, as per EU definitions), in many cases, 'new entrants' to farming, who formed niche tandems for innovation, were returning residents: individuals who were raised in the region but left to pursue employment elsewhere, returning with new business skills and ideals of rural life (for example, in the Portuguese lifestyle farming case and the Greek local quality convention, Karditsa). These individuals re-entered agriculture as 'hybrid actors', enabling niche development. These actors appear more open to innovation, and to establishing partnerships to undertake activities outside of normal production processes, challenging the incumbent regime. The examples above show that newcomers to the agricultural sector do not necessarily have to be farmers but are often members of other sectors, or have different backgrounds.

It was also evident that a particular type of innovation does not always emerge from the same type of source. For example, Darrot *et al.* (this volume) demonstrate that AAFN were initiated by farmers in France, consumers in the Czech Republic and retailers from the tourist industry in Greece. This is consistent with the tenet in the MLP that niches arise on the periphery, or in another regime. Therefore, in seeking to support agricultural innovation and transition, governments should not necessarily look solely to the agricultural sector. Interventions could also usefully encourage young people from rural areas to experience urban employment and return to rural regions later in life.

Enabling sustainability transitions in the agricultural sector

In this section, we reflect on findings in relation to the types of transition processes (towards productive, consumptive and protective functions) the cases in this book inform, whilst recalling Pinto-Correia *et al.*b's (this volume) finding that regions typically aspire to a combination of functions for their agricultural sectors.

The research findings suggest that the MLP is well suited to understanding initiatives focused on altering the *productive function* of agriculture. Concepts were easily implemented in the study of on-farm renewable energy production (Sutherland *et al.*b, this volume): a clear technological basis led to a commercial product, which fitted well with existing farming production norms and benefited from both state and industry investment. The findings suggest that such production-oriented transitions may be the easiest to mainstream. However, although anchoring into a new production-based regime can provide a source of stability for individual farms, the focus on efficient production in the agricultural sector as a whole, has arguably led to many of the problems being addressed by other sustainability initiatives (such as environmental risks related to intensive farming, Vlahos and Schiller, this volume; negative local economic effects of globalized food production, Darrot *et al.* and Lošťák *et al.*, both this volume). Although renewable energy production represents decentralization in the energy sector, it may also contribute to increased concentration and intensification in the agricultural sector because it tends to be located on large-scale farms which can leverage the considerable investment required. Farm-based renewables also face considerable competition from non-agricultural businesses interested in capitalizing on energy markets, and there is increasing public concern about the secondary impacts of renewable energy production, particularly the mono-cropping of maize for biogas plants and the visual impact of turbines on the surrounding countryside. Pursuit of production-based pathways in isolation can, thus, be highly contentious, risking exacerbation of existing sustainability issues.

Local certification and AAFN also produce products – food – but in opposition to existing marketing channels. In doing so, they typically do not intend to change the existing markets; instead they aim to be autonomous through the developments of new markets. Whilst this strategy has been successful in the cases studied, there is increasing evidence that these emerging markets are targeted by multinational grocery chains who are similarly integrating 'local' products into their marketing strategies. These niches are, thus, influencing incremental regime change within the agri-food sector but in so doing, may limit their own long-term viability. These power dynamics within multi-regime interactions are an important area for future research.

Local certification channels and AAFN also demonstrate cases where the consumption function of agriculture is promoted (such as the creation of food products that appeal to consumers on the basis of heritage, quality and other amenity features). Although these produce marketable products and, as such, are recognized by the policy sector, the utility of the MLP in assessing these types of transitions is less evident. Lifestyle farming, in particular, demonstrates that consumers can initiate changes within agriculture but the absence of clear products makes it much more difficult to define the niche and its associated interactions. Network anchoring appears to be largely absent for lifestyle farmers, and Pinto-Correia *et al.*a (this volume) describe this new type of farmer as 'unseen', owing to their lack of state recognition, regulation and integration into agricultural knowledge networks. However, the consumption function of agriculture appears to be directly linked to hybridization, establishing connections between multiple regimes. New entrants to farming and the agricultural sector bring with them new visions of the purpose and potential of farming activities, as well as knowledge and skills from other sectors. The resultant 'niche-tandems', 'retro-innovations' and 'hybrid markets' are important sources of innovation for farming. However, the development of this consumption function raises the question of who consumption is for (in other words, only those who can afford specialized products or services), as well as increasing competition for rural resources such as land. We suggest that the development of this consumption function should not be ignored, nor should it be treated as a means of increasing average levels of income in rural areas without recognizing the underlying power dynamics (the winners, losers, and long term implications for resource access).

In contrast, support for the *protection function* of agriculture is where multifunctional agricultural policy – at least in northern Europe – has focused in recent decades (Wilson, 2007). The natural resource initiatives discussed in this book were established to counter water quality issues resulting from intensive agricultural production (Vlahos and Schiller, this volume), whereas HNVF is about the social, economic and environmental sustainability of remote, marginal agricultural areas (Peneva *et al.*, this volume). Smith (2007) points out the challenges in establishing green niches, arguing that linking to the regime is easier for niches which are aligned to it but the closer the alignment, the less likely it is that the niches will lead to a radical transition. In their French case study, Darrot *et al.* (this volume) describe how, despite the development of a well-anchored niche (including low input farmers, suppliers, scientists, tourist industry members and government officials) the political pressure brought about by intensive farmers led to weak mediation strategies and, ultimately, the failure of government policies to eradicate toxic 'green tides' caused by water pollution. In contrast, the HNVF areas were very successful in engaging new participants, including young people. The challenges for supporting protection functions appear to be far greater in regions where land is more productive (where market forces support the production function). The challenge of long-term funding for these public goods is also an issue; although the research has identified some promising options (such as shareholder cooperatives and payments for ecosystem services) the variable market value of different social and environmental services, within a context of increasing scarcity of state resources, raises the questions of the winners and losers in these types of funding strategies.

In identifying particular trajectories for agriculture, it is important to note that the scaling-up of innovations was not necessarily the intention of the initiatives studied. Proliferation, when it occurred, was more viable (as local food, by definition, can only be local to a specified geographical area). Innovations adopted in the regions studied were also not necessarily local to that region. For example, the machinery ring concept was already successful in Germany before it was introduced to Scotland by state-funded agricultural advisors. The CRIE Montado niche was established to follow a Dutch programme. The renewable energy technologies were largely developed outside of the regions that were studied, and then imported by enthusiastic actors. Further, local certification schemes and AAFN may have been pioneered in the French cases but it is clear that founders of more recently established cases, in the Czech Republic and Bulgaria, had observed similar processes elsewhere. In addition, as Darnhofer *et al.* (this volume) outlined, one niche can affect the regime, paving the way for others. Smith and Raven (2012) described this as a change in the selection context ('stretch and transform empowerment'), where the regime is transformed through institutionalization of some features of the niche. Successful development may, thus, result from adoption of practices which have been successful elsewhere, rather than new innovation *per se*.

Future directions and looking forward

The FarmPath project was developed in response to a European Commission call in 2009, for research on enabling sustainability transitions in agriculture. Since then, concerns regarding food security and the demands of an increasing world population have shifted policy discourse. Whereas Slee and Pinto-Correia (this volume) demonstrated the policy shifts towards protection and consumption functions in European rural areas over the past three decades, new terms like 'sustainable intensification' and support for the 'bio-economy' have recently permeated EU research calls. It is now generally acknowledged that productivist agricultural and rural pathways have continued unabated in most areas of the world, and some are even beginning to refer to 'neo-productivist' pathways, sparked by demands for further intensification of agricultural spaces through global population growth; the use of food crops for production of biofuels; changing diets towards meat and dairy products in transition economies; and diminishing yields in parts of the world linked to climate change (Burton and Wilson, 2012; Perkins, 2012). These processes will continue to influence future transitions towards the sustainability of the European countryside, in particular where they exacerbate the competition between production, consumption and protection functions in rural spaces.

In the light of this policy shift towards the productive function of agriculture, it is particularly important to continue to develop the MLP for use in relation to agricultural-sector transitions. The findings from this research demonstrate the utility of the MLP for assessing production-oriented transition processes, as well as drawing attention to the role of enabling technologies such as IT. The empirical application of the MLP to emerging transition processes has also revealed a number of important concepts that will benefit from ongoing research. In particular, *hybridity*, resulting from cross-sectoral interaction, was found to be central to the processes of transition, leading to the formation of niche-tandems, hybrid actors and markets, retro-innovation, and resulting in regime graffiting of external policy structures. The findings, thus, demonstrate the need for further development of *multiple regime interaction* within the MLP, in order to better understand the development of non-technological transition processes. To date, the few studies which exist typically address interactions between regimes defined within a single sector (e.g. Geels, 2007; Raven and Verbong, 2007; Konrad *et al.*, 2008); whereas the multifunctional nature of the agricultural sector lends itself particularly to cross-sectoral development. As such, MLP research could also usefully be integrated with the substantial body of literature on multifunctional agriculture in Europe, and rural development more broadly (through concepts of social capital, social exclusion and regional development). These literatures, in turn, would benefit from the utility of the MLP for addressing change at multiple levels, and for highlighting the role of technology in transition processes.

There are also important issues of *space and scale* that need to be considered in future research on sustainability transitions. Our analysis has demonstrated the regional specificity of such transitions, and the need to locate them in specific geographical locations (in contrast to MLP approaches which tend to follow the development of technology a-spatially). Many of the cases studied represent the proliferation of innovations developed elsewhere, and were influenced by transition processes and technological developments from outside of the agricultural sector. These *secondary transition processes* require further conceptual development, particularly for the agricultural sector where, as Darnofer *et al.* (this volume) note, radical transition is less likely owing to the complexity of the system.

Finally, *power relations* require further development within the MLP, particularly as they relate to natural resource access and the protection and consumption functions of agriculture. The sovereignty of market forces is largely unchallenged within the MLP, leaving unaddressed the social justice issues associated with market-based outcomes. The role of consumers, particularly in relation to the consumption function of agriculture, merits particular attention. Access to healthy food, energy, housing and recreation are all key functions of the agricultural sector, which must be balanced in the pursuit of sustainability transitions.

References

Belz, F.M. (2004) A transition towards sustainability in the Swiss agri-food chain (1970-2000): Using and improving the multi-level perspective. In: Elzen, B., Geels, F.W. and Green, K. (eds) *System Innovation and the Transition to Sustainability.* Edward Elgar Publishing, Cheltenham, UK, pp. 97–113.

Burton, R.J.F. and Wilson, G.A. (2012) The rejuvenation of productivist agriculture: The case for 'cooperative neo-productivism'. In: Almås, R. and Campbell, H. (eds) *Rethinking Agricultural Policy Regimes: Food Security, Climate Change and the Future Resilience of Global Agriculture (Research in Rural Sociology and Development, Volume 18).* Emerald Group Publishing Limited, Bingley, UK, pp. 51–72.

Darnhofer, I., Sutherland, L-A. and Pinto-Correia, T. (2015) Conceptual insights derived from case studies on 'emerging transitions' in farming. In: Sutherland, L-A., Darnhofer, I., Wilson, G.A. and Zagata, L. (eds) *Transition Pathways towards Sustainability in Agriculture: Case Studies from Europe.* CABI, Wallingford, UK, pp. 189–204.

Darrot, C., Diaz, M., Tsakalou, E. and Zagata, L. (2015) 'The missing actor': Alternative agri-food networks and the resistance of key regime actors. In: Sutherland, L-A., Darnhofer, I., Wilson, G.A. and Zagata, L. (eds) *Transition Pathways towards Sustainability in Agriculture: Case Studies from Europe.* CABI, Wallingford, UK, pp. 143–156.

DGIP (2012) EU Measures to encourage and support new entrants. Note published by Directorate-general for internal policies. Available at: http://www.europarl.europa.eu/committees/en/agri/

Elzen, B., van Mierlo, B. and Leeuwis, C. (2012) Anchoring of innovations: Assessing Dutch efforts to harvest energy from glasshouses. *Environmental Innovations and Societal Transitions* 5, 1–18.

Geels, F.W. (2004) From sectoral systems of innovation to socio-technical systems. Insights about dynamics and change from sociology and institutional theory. *Research Policy* 33, 897–920.

Geels, F.W. (2007) Analysing the breakthrough of rock 'n' roll (1930–1970). Multi-regime interaction and reconfiguration in the multi-level perspective. *Technological Forecasting and Social Change* 74, 1411–1431.

Geels, F.W. and Schot, J. (2007) Typology of sociotechnical transition pathways. *Research Policy* 26, 399–417.

Geels, F.W. and Schot, J. (2010) The dynamics of transitions. A socio-technical perspective. In: Grin, J., Rotmans, J. and Schot, J. (eds) *Transitions to Sustainable Development.* Routledge, New York, pp. 11–101.

Karanikolas, P., Vlahos, G. and Sutherland, L-A. (2015) Utilising the multi-level perspective in empirical field research: Methodological considerations. In: Sutherland, L-A., Darnhofer, I., Wilson, G.A. and Zagata, L. (eds) *Transition Pathways towards Sustainability in Agriculture: Case Studies from Europe.* CABI, Wallingford, UK, pp. 51–66.

Konrad, K., Truffer, B. and Voß, J-P. (2008) Multi-regime dynamics in the analysis of sectoral transformation potentials: Evidence from German utility sectors. *Journal of Cleaner Production* 16, 1190–1202.

Lošťák, M., Karanikolas, P., Draganova, M. and Zagata, L. (2015) Local quality and certification schemes as new forms of governance in sustainability transitions. In: Sutherland, L-A., Darnhofer, I., Wilson, G.A. and Zagata, L. (eds) *Transition Pathways towards Sustainability in Agriculture: Case Studies from Europe.* CABI, Wallingford, UK, pp. 157–170.

Markhard, J. and Truffer, B. (2008) Technological innovation systems and the multi-level perspective: Towards an integrated framework. *Research Policy* 37, 596–615.

Marsden, T. (2013) From post-productionism to reflexive governance: Contested transitions in securing more sustainable food futures. *Journal of Rural Studies* 29, 123–134.

Peneva, M., Draganova, M., Gonzalez, C., Diaz, M. and Mishev, P. (2015) High nature value farming: Environmental practices for rural sustainability. In: Sutherland, L-A., Darnhofer, I., Wilson, G.A. and Zagata, L. (eds) *Transition Pathways towards Sustainability in Agriculture: Case Studies from Europe.* CABI, Wallingford, UK, pp. 97–112.

Perkins, C. (2012) Neo-productivism, farm families, and technologies in Dundee, UK. In: Harakova, H. and Boscoboinink, A. (eds) *From Production to Consumption: Transformation of Rural Communities in Europe.* LIT Verlag, Munich, Germany, pp. 165–180.

Pinto-Correia, T., Gonzalez, C., Sutherland, L-A. and Peneva, M. (2015a) Lifestyle farming: Countryside consumption and transition towards new farming models. In: Sutherland, L-A., Darnhofer, I., Wilson, G.A. and Zagata, L. (eds) *Transition Pathways towards Sustainability in Agriculture: Case Studies from Europe.* CABI, Wallingford, UK, pp. 67–82.

Pinto-Correia, T., McKee, A. and Guimarães, H. (2015b) Transdisciplinarity in deriving transition pathways for agriculture. In: Sutherland, L-A., Darnhofer, I., Wilson, G.A. and Zagata, L. (eds) *Transition Pathways towards Sustainability in Agriculture: Case Studies from Europe.* CABI, Wallingford, UK, pp. 171–188.

Pretty, J. (2003) Social capital and the collective management of resources. *Science* 302, 1912–1914.

Raven, R. and Verbong, G. (2007) Multi-regime interactions in the Dutch energy sector: The case of combined heat and power technologies in the Netherlands 1970 – 2000. *Technology Analysis & Strategic Management* 19, 491–507.

Rip, A. and Kemp, R. (1998) Technological change. In: Rayer, S. and Malone, E.L. (eds) *Human Choice and Climate Change.* Battelle Press, Columbus, Ohio.

Schiller, S.R., Gonzalez, C. and Flanigan, S. (2015) More than just a factor in transition processes? The role of collaboration in agriculture. In: Sutherland, L-A., Darnhofer, I., Wilson, G.A. and Zagata, L. (eds) *Transition Pathways towards Sustainability in Agriculture: Case Studies from Europe.* CABI, Wallingford, UK, pp. 83–96.

Slee, B. and Pinto-Correia, T. (2015) Understanding the diversity of European rural areas. In: Sutherland, L-A., Darnhofer, I., Wilson, G.A. and Zagata, L. (eds) *Transition Pathways towards Sustainability in Agriculture: Case Studies from Europe.* CABI, Wallingford, UK, pp. 33–50.

Smith, A. (2006) Green niches in sustainable development: The case of organic food in the United Kingdom. *Environment and Planning C: Government and Policy* 24, 439–458.

Smith, A. (2007) Translating sustainabilities between green niches and socio-technical regimes. *Technology Analysis & Strategic Management* 19, 427–450.

Smith, A. and Raven, R. (2012) What is protective space? Reconsidering niches in transitions to sustainability. *Research Policy* 41, 1025–1036.

Sutherland, L-A. and Holstead, K. (2014) Future-proofing the farm: On-farm wind turbine development in farm business decision-making. *Land Use Policy* 36, 102–112.

Sutherland, L-A., Peter, S. and Zagata, L. (2015b) On-farm renewable energy: A 'classic case' of technological transition. In: Sutherland, L-A., Darnhofer, I., Wilson, G.A. and Zagata, L. (eds) *Transition Pathways towards Sustainability in Agriculture: Case Studies from Europe.* CABI, Wallingford, UK, pp. 127–142.

Sutherland, L-A., Wilson, G.A. and Zagata, L. (2015a) Introduction. In: Sutherland, L-A., Darnhofer, I., Wilson, G.A. and Zagata, L. (eds) *Transition Pathways towards Sustainability in Agriculture: Case Studies from Europe.* CABI, Wallingford, UK, pp. 1–16.

Vlahos, G. and Schiller, S. (2015) Transition processes and natural resource management. In: Sutherland, L-A., Darnhofer, I., Wilson, G.A. and Zagata, L. (eds) *Transition Pathways towards Sustainability in Agriculture: Case Studies from Europe.* CABI, Wallingford, UK, pp. 113–126.

Wilson, G.A. (2007) *Multifunctional Agriculture: A Transition Theory Perspective*. CABI, Wallingford, UK.

Wilson, G.A. (2010) Multifunctional 'quality' and rural community resilience. *Transactions of the Institute of British Geographers* 35, 364–381.

Wilson, G.A. (2012) *Community Resilience and Environmental Transitions*. Routledge/Earthscan, London, UK.

A

B

C

Q

R

S

T

U

V